Tissue Phenomics

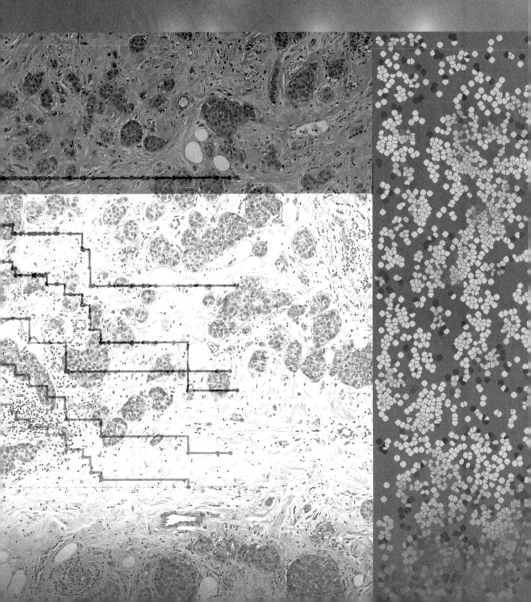

Tissue Phenomics

Profiling Cancer Patients for Treatment Decisions

edited by

Gerd Binnig
Ralf Huss
Günter Schmidt

PAN STANFORD PUBLISHING

Published by

Pan Stanford Publishing Pte. Ltd.
Penthouse Level, Suntec Tower 3
8 Temasek Boulevard
Singapore 038988

Email: editorial@panstanford.com
Web: www.panstanford.com

British Library Cataloguing-in-Publication Data
A catalogue record for this book is available from the British Library.

**Tissue Phenomics: Profiling Cancer Patients for
Treatment Decisions**
Copyright © 2018 by Pan Stanford Publishing Pte. Ltd.
*All rights reserved. This book, or parts thereof, may not be reproduced in any form
or by any means, electronic or mechanical, including photocopying, recording
or any information storage and retrieval system now known or to be invented,
without written permission from the publisher.*

For photocopying of material in this volume, please pay a copying fee through
the Copyright Clearance Center, Inc., 222 Rosewood Drive, Danvers, MA 01923,
USA. In this case permission to photocopy is not required from the publisher.

ISBN 978-981-4774-88-8 (Hardcover)
ISBN 978-1-351-13427-9 (eBook)

Morphological and functional complexity of animal and human tissue is one of the most fascinating and simultaneously challenging topics in tissue-based science. The diverse organizational units of a normal liver, lung, or kidney, such as organ-specific cells, nerves, connective tissues, and different types of vessels, show so many variants that a systematic and comprehensive analysis by human eyes is not possible. In addition, time-related modifications and spatial distribution of the components as well as disease-related variants produce an even higher level of complexity, often termed hyper-complexity.

Today, the only way to find a solution for reliable and reproducible analyses of various tissues is based on multiplex digital systems that produce tissue-related big data. By applying suitable algorithms, these data can be sorted and used to answer different diagnostic, prognostic, predictive, and scientific questions. The concept of tissue phenomics is currently the most promising approach to answer many burning questions of cancer and other diseases.

—Prof. Dr. med. Dr. h. c. Manfred Dietel

Charité, Berlin, Germany

Contents

Foreword	xiii
Acknowledgments	xvii

1. Introduction to Tissue Phenomics — **1**

Ralf Huss, Gerd Binnig, Günter Schmidt,
Martin Baatz, and Johannes Zimmermann

1.1	Motivation	1
1.2	Tissue Phenomics and Medicine 4.0	3
1.3	Phenes in Cancer Immunology	5
1.4	Future of Tissue Phenomics	6
1.5	About the Book	7

2. Image Analysis for Tissue Phenomics — **9**

Johannes Zimmermann, Keith E. Steele, Brian Laffin,
René Korn, Jan Lesniak, Tobias Wiestler, and
Martin Baatz

2.1	Introduction	10
2.2	Experimental Design for Image Analysis Studies	12
2.3	Input Data and Test Data	14
2.4	Image Analysis	15
2.5	Quality Control Procedures	25
2.6	Results and Output	30
2.7	Discussion and Summary	31

3. Context-Driven Image Analysis: Cognition Network Language — **35**

Gerd Binnig

3.1	Motivation and Reasoning		36
3.2	History and Philosophy		40
3.3	Technology		43
	3.3.1	Class Hierarchy	44
	3.3.2	Process Hierarchy and Context Navigation	44

		3.3.2.1	Processes and the domain concept	44
		3.3.2.2	Context navigation	45
		3.3.2.3	Object-based processing	46
		3.3.2.4	Maps	48
		3.3.2.5	Object variables	49
3.4	Example of Context-Driven Analysis			50
	3.4.1	H&E Analysis Problems		53
	3.4.2	Concrete H&E Image Analysis Example		54
		3.4.2.1	Color channel analysis as a starting point	54
		3.4.2.2	Isolated nuclei: first context objects	55
		3.4.2.3	Some remaining large nuclei by splitting and *inside-out* procedure	60
		3.4.2.4	Nuclei of immune cells	61
		3.4.2.5	Template match for heterogeneous nuclei on 10× and 20×	63
3.5	Conclusion			65

4. Machine Learning: A Data-Driven Approach to Image Analysis **69**

Nicolas Brieu, Maximilian Baust, Nathalie Harder, Katharina Nekolla, Armin Meier, and Günter Schmidt

4.1	Introduction: From Knowledge-Driven to Data-Driven Systems			70
4.2	Basics of Machine Learning			71
	4.2.1	Supervised Learning		72
	4.2.2	Classification and Regression Problems		72
	4.2.3	Data Organization		73
4.3	Random Forests for Feature Learning			77
	4.3.1	Decision Trees		77
		4.3.1.1	Definition	77
		4.3.1.2	Decision function	78
		4.3.1.3	Extremely randomized trees	78
		4.3.1.4	Training objective function	79
	4.3.2	Random Forests Ensemble (Bagging)		79
	4.3.3	Model Parameters		80

	4.3.4	Application Generic Visual Context Features	81
		4.3.4.1 Haar-like features	81
		4.3.4.2 Gabor features	82
	4.3.5	Application to the Analysis of Digital Pathology Images	82
		4.3.5.1 On-the-fly learning of slide-specific random forests models	83
		4.3.5.2 Area-guided distance-based detection of cell centers	84
4.4	Deep Convolutional Neural Networks		86
	4.4.1	History: From Perceptrons and the XOR Problem to Deep Networks	87
	4.4.2	Building Blocks	89
		4.4.2.1 Convolutional layers	89
		4.4.2.2 Convolutional neural networks	90
		4.4.2.3 Loss functions	91
		4.4.2.4 Activation functions	91
		4.4.2.5 Pooling layers	92
		4.4.2.6 Dropout layers	93
	4.4.3	Application Examples	94
4.5	Discussion and Conclusion		97
	4.5.1	Model Complexity and Data Availability	97
	4.5.2	Knowledge and Data Teaming	97
	4.5.3	Machine Learning Teaming	98
	4.5.4	Conclusion	98

5. Image-Based Data Mining   **101**

Ralf Schönmeyer, Arno Schäpe, and Günter Schmidt

5.1	Introduction		102
5.2	Generating High-Level Features for Patient Diagnosis		104
	5.2.1	Quantification of Regions of Interest	104
	5.2.2	Heatmaps	105
		5.2.2.1 Contents of heatmaps	106
		5.2.2.2 Visualization of heatmaps	107
		5.2.2.3 Heatmaps with multiplexed data	108

| | | 5.2.2.4 Objects in heatmaps | 110 |

	5.2.3	Algebraic Feature Composition	111
	5.2.4	Aggregation of Multiple Tissue Pieces to Single Patient Descriptors	113
	5.2.5	Integration of Clinical Data and Other Omics Data	113
5.3	Performance Metrics		115
5.4	Feature Selection Methods		117
	5.4.1	Unsupervised Methods	117
	5.4.2	Hypothesis-Driven Methods	118
	5.4.3	Univariate, Data-Driven Feature Selection	120
5.5	Tissue Phenomics Loop		121
5.6	Tissue Phenomics Software		123
	5.6.1	Image Analysis and Data Integration	123
	5.6.2	General Architecture	123
	5.6.3	Image Mining Software	124
	5.6.4	Data and Workflow Management and Collaboration	125
5.7	Discussion and Outlook		127

6. Bioinformatics   131

Sriram Sridhar, Brandon W. Higgs, and Sonja Althammer

6.1	Molecular Technologies: Past to Present		132
6.2	Genomics and Tissue Phenomics		134
	6.2.1	Genomics Data Sources	134
	6.2.2	The Art of Image Mining	138
		6.2.2.1 Cell-to-cell distances	139
		6.2.2.2 Quantifying cell populations in histological regions	140
		6.2.2.3 Tissue phene and survival	142
	6.2.3	Power of Integrative Approaches	142
6.3	Analytical Approaches for Image and Genomics Data		144
	6.3.1	Data Handling Requires Similar Methods	144
	6.3.2	Enhancing Confidence in a Discovery	145
6.4	Examples of Genomics or IHC Biomarkers in Clinical Practice		147
	6.4.1	Biomarker Background	147

	6.4.2	Genomics and IHC to Guide Prognosis or Diagnosis	148
	6.4.3	Patient Stratification: Genomics and IHC to Identify Patient Subsets for Treatment	150

7. Applications of Tissue Phenomics

157

Johannes Zimmermann, Nathalie Harder, and Brian Laffin

7.1	Introduction		158
7.2	Hypothesis-Driven Approaches		159
	7.2.1	TME as a Battlefield: CD8 and PD-L1 Densities Facilitate Patient Stratification for Durvalumab Therapy in Non–Small Cell Lung Cancer	159
	7.2.2	Gland Morphology and TAM Distribution Patterns Outperform Gleason Score in Prostate Cancer	159
	7.2.3	Novel Spatial Features Improve Staging in Colorectal Cancer	160
	7.2.4	Immunoscore Is a Novel Predictor of Patient Survival in Colorectal Cancer	162
	7.2.5	Analysis of Spatial Interaction Patterns Improves Conventional Cell Density Scores in Breast Cancer	163
	7.2.6	Immune Cell Infiltration is Prognostic in Breast Cancer	165
	7.2.7	The Immune Landscape Structure Directly Corresponds with Clinical Outcome in Clear Cell Renal Cell Carcinoma	166
7.3	Hypothesis-Free Prediction		166
7.4	Summary		167

8. Tissue Phenomics for Diagnostic Pathology

175

Maria Athelogou and Ralf Huss

8.1	Introduction	175
8.2	Digital Pathology	176
8.3	Tissue Phenomics Applications for Computer-Aided Diagnosis in Pathology	178

| | 8.3.1 | Technical Prerequisites for Tissue Phenomics | 180 |

8.3.1 Technical Prerequisites for Tissue Phenomics — 180

8.4 Tissue Phenomics Applications for Decision Support Systems in Pathology — 180

8.5 Summary of Pathology 4.0 — 182

9. Digital Pathology: Path into the Future — 185

Peter D. Caie and David J. Harrison

9.1 Introduction — 186

9.2 Brief History of Digital Pathology — 187

9.3 Digital Pathology in Current Practice — 188

    9.3.1 Research — 188

    9.3.2 Education — 189

    9.3.3 Clinical — 190

9.4 Future of Digital Pathology — 191

    9.4.1 Mobile Scanning — 191

    9.4.2 Feature-Based Image Analysis — 192

    9.4.3 Machine Learning on Digital Images — 195

    9.4.4 Big Data and Personalized Pathology — 196

10. Tissue Phenomics in Clinical Development and Clinical Decision Support — 199

Florian Leiß and Thomas Heydler

10.1 Cancer and Oncology Drug Development — 199

10.2 Immunotherapy — 200

10.3 Digitalization of Healthcare — 202

10.4 Patient Profiling — 203

10.5 Therapy Matching — 205

10.6 Benefits — 207

10.7 Conclusion — 208

Glossary — 209

Index — 215

Foreword

Evolution of Tissue Phenomics and
Why It Is Critical to the War on Cancer

A view from a tumor immunologist
and cancer immunotherapist

Those of us in biomedical research are witnessing an almost daily evolution of our science. Nowhere is this more obvious, or possessing greater impact, than in the field of cancer immunology and immunotherapy. Cancer, one of the great scourges on humanity, is having the veil of its secrets lifted. Digital imaging and objective assessment tools contribute substantial and solid evidence to document that immune cells are prognostic biomarkers of improved outcomes for patients with cancer. While anecdotal reports of associations between immune cell infiltrates and improved outcomes have been presented by pathologists for more than 100 years, the co-evolution of multiple science subspecialties has resulted in opportunities to better understand the disease and why it develops. Armed with this knowledge and evidence that checkpoint blockade therapies are capable of unleashing the immune system, increasing survival and possibly curing some patients with cancer, will lead to additional investment in this area of research, which will accelerate the pace at which we develop improved treatments for cancer. It is clear that digital imaging and assessment of complex relationships of cells within cancer, the very essence of tissue phenomics will play a central role in the development of the next generation of cancer immunotherapies. Ultimately, assessment of cancer tissue phenomics will be used to tailor immunotherapies to treat and eventually cure patients with cancer.

This is a very different time from when I began as an immunologist. Monoclonal antibodies, reagents capable of objectively assessing thousands of molecules, were not yet invented. To identify a subset of white blood cells, termed T cells, we used the binding of sheep erythrocytes to lymphocytes as the assay to characterize their numbers. The number of lymphocytes that formed *rosettes* was counted on a hemocytometer using a light microscope. This method was used to dose patients with immunosuppressive therapy to prevent allograft rejection. In tissue sections and smears, we did not use immunohistochemistry (IHC); we used ocular annotation of a cell's morphological characteristics. My limited training in this area came at the hands of Dr. John W. Rebuck, a hematopathologist, who trained with Dr. Hal Downey, who trained with Professor Artur Pappenheim, at the University of Berlin. Professor Pappenheim, who developed the Pappenheim stains, educated his trainees into the subtleties of morphology, who then propagated the method to their trainees, and in this way the method spread.

Lymphocyte was my cell of interest; it was known to be a small round cell, with limited cytoplasm. Under Dr. Rebuck's tutelage, I denuded my skin with a scalpel, placed a drop of the diphtheria, pertussis, and tetanus vaccine on the area, covered it with a sterile glass coverslip, and attached it in place using a small piece of cardboard and adhesive tape. I would then change the coverslips every 3 h over a period of 48 h. Once the coverslips were removed and stained with Leishman's, I would sit at a multi-headed microscope and Dr. Rebuck would point out the monocytes or lymphocytes that were migrating into the site and onto the coverslip, identifying whether they were lymphocytes or monocytes that were morphing into large phagocytic cells. To support his description, Dr. Rebuck would describe the nuclear and cytoplasmic characteristics, as well as the other types of cells present in the area. That type of detailed evaluation has gone on for more than a century and remains the principal means to characterize disease.

As you read this book, it will become clear how the advances in image assessment technology allow subtle characteristics of cells to be evaluated in an objective and automated fashion. In addition to the morphological characteristics of cells, multiplex IHC provides simultaneous assessment of six or more markers on a single slide. Utilizing other technology, it is possible to stain a slide and then image and strip the slide of the reagents so that the cycle can be repeated as many as 60 times, allowing assessment of as many markers on a single slide. Coupled with the advent of tissue phenomics, all of this information can be evaluated in the context of whether it is inside the tumor, at the invasive margin, or in the stroma.

The molecular evaluation of disease is also advancing. Summaries of gene expression profiling data for tumor samples from hundreds of patients are available in The Cancer Genome

Atlas (TCGA) and other databases. These databases are being interrogated to evaluate how many cancers had high or low levels of genes associated with immune cells, as well as for expression of cancer-destroying molecules or of mechanisms cancer can use to evade immune-mediated destruction. While these interrogations of the data are providing important and powerful insights about the immune system's response to cancer, much of this information lacks the context of where these elements are expressed and the identity of which cells are expressing specific genes. Only by understanding the context of this information, specifically which cell is expressing what and which cells are nearby, can it be effectively used to guide a new generation of cancer immunotherapy trials.

Head and neck cancer will serve as an example to further clarify this point. For several years, it has been known that increased numbers of CD8(+) *cancer killer T cells* at the tumor were associated with improved survival. For example, in one study patients whose tumors had above the median number of CD8 T cells had around a 50% 5-year survival, while those whose tumors had CD8 T cell numbers below the median, had a 35% 5-year survival. This pattern was true for assessment of CD8(+) T cell numbers by IHC or gene expression profiling. As an immunologist who recognizes the important role that CD8 T cells can play in preclinical animal models, this made sense. However, it was also reported that an increased number of FOXP3(+) *suppressor* T cells were also associated with improved survival. Since these are the cells that can turn off the cancer killer cells, these results made no sense. As we began to apply the multiplex IHC method to visualize six markers on a single 4-micron section, it became clear that in some patients, immune cells associated with the tumor were organized in a specific pattern. In one case, the tumor, which uniformly expressed high levels of the immune checkpoint PD-L1, had excluded essentially all immune cells from inside the tumor. However, at the tumor–stroma barrier was a band of suppressor T cells, and outside of that band were the CD8(+) cancer killer T cells. Since several of the suppressor cells' immune inhibitory functions required cell contact, we reasoned that in order to be effective, the suppressor cells would need to be relatively close to the CD8(+) cancer killer cell that they were trying to inhibit. When we evaluated tumors with a high number of suppressor cells near the CD8(+) cancer killer cells, we found that these patients did significantly worse than patients with low numbers, not better. As we evaluated another inhibitory molecule, PD-L1, which mediates inhibition by contact, we found the same pattern. While additional validation needs to be done, this illustrates the power of evaluating cell–cell relationships. It will be these types of assessments with panels of 20–25 markers that will provide critical insights into why tumors escape immune elimination and what hurdles will need to be addressed to improve patient outcomes. It is not going to be easy and after spending 5 years in the midst of digital imaging technologies, I realize there are substantial challenges ahead. There is no looking back! Since the immune system is the critical element that can cure patients of cancer, it is essential to assess the cancer that escapes and ultimately kills the patient. While multiple approaches must be applied, the *tissue is the issue* and tissue phenomics is the approach that will play the critical role in unraveling the amazing complexity of the cancer–immune system interphase and drive the development of treatments to cure patients with cancer.

Bernard A. Fox
Providence Cancer Center
Portland, OR, USA
Autumn 2017

Acknowledgments

We express our gratitude to all co-authors and contributors, who worked with great commitment and passion on the successful completion of the different chapters and the finishing of the entire book.

We also thank everybody in Definiens AG for insightful and astute discussions to provide a comprehensive and coherent overview on *tissue phenomics* and related subjects, such as cognition networks, assay development, digital pathology, image analysis, big data analytics, and machine learning. We very much appreciate the generous support of Definiens AG in providing us leeway and resources to create this book.

A very special thank goes to Ralf Schönmeyer for his enthusiastic organizational support, manuscript editing, and putting everything together in the desired format.

We thank the members of our advisory board for their devoted leadership and inspirational guidance and our investors and shareholders for their solid confidence and trust in such a disruptive technology and our creative people.

We are grateful to our longtime academic collaborators, business partners, and commercial customers for their outstanding insights and challenging thoughts. This holds true in particular for the attendees of our annual *Tissue Phenomics Symposium* and their enthusiasm for novel ideas and creative solutions.

And finally, we thank Pan Stanford Publishing for inviting us to write this book, contribute to the editorial work, and for subsequently publishing it.

Gerd Binnig
Ralf Huss
Günter Schmidt

Chapter 1

Introduction to Tissue Phenomics

Ralf Huss,[a] Gerd Binnig,[b] Günter Schmidt,[a] Martin Baatz,[a] and Johannes Zimmermann[a]

[a]*Definiens AG, Bernhard-Wicki-Strasse 5, 80636 Munich, Germany*
[b]*Former CTO and founder of Definiens AG, Bernhard-Wicki-Strasse 5, 80636 Munich, Germany*
rhuss@definiens.com

1.1 Motivation

For many decades, tumor tissue has been the basis and gold standard for cancer diagnosis and treatment, putting the pathologist close to patient care by giving essential guidance for treatment decisions. Extracting information from tissue in histopathology often is still predominantly a manual process performed by expert pathologists. This process, to a certain degree, is subjective and depends on the personal experience of each pathologist; rather qualitative than quantitative and of limited repeatability as the considerable variability in histological grading among pathologists exemplifies. Moreover, conventional histopathology currently does not allow systematic extraction of the rich information residing in tissue sections. The histological section, being a two-dimensional representation of three-dimensional structures, represents the complex phenotype in a solid way. The entirety of DNA, RNA, proteome, and metabolome

Tissue Phenomics: Profiling Cancer Patients for Treatment Decisions
Edited by Gerd Binnig, Ralf Huss, and Günter Schmidt
Copyright © 2018 Pan Stanford Publishing Pte. Ltd.
ISBN 978-981-4774-88-8 (Hardcover), 978-1-351-13427-9 (eBook)
www.panstanford.com

are all blended into a topologically ordered and preserved morphological integrity. Therefore, the transition from a more descriptive to a strictly quantitative discipline, being the core objective of modern histopathology, is clearly worthwhile to pursue. This will allow the extraction of large amounts of tissue-derived data from many patients, enabling the creation of unprecedented knowledge through statistical evaluations and improving treatment of individual patients.

Conventional approaches such as H&E-based morphological assessment have been increasingly complemented by proteomic and genetic methods, mainly through immunohistochemistry (IHC) and *in situ* hybridization. Genomic studies, although not providing microscopic spatial resolution, started to add value by furnishing rich genetic fingerprints. Several successful attempts have been developed to understand cancer determinants in tissue-based genomics with platform technologies such as qPCR and next-generation sequencing (NGS). The importance of genomics in oncology is natural and appears evident as cancer can be a genetic disease driven by specific genetic mutations. In a number of cases, correlations between gene mutations and diseases have been identified, and associated diagnostic tests have been established. However, today only limited parts of available data are of known clinical relevance and are actually used for diagnostic or therapeutic decision-making. Tumors are known to be heterogeneous and many different mutations or wild-type gene expression-related features occur that might affect patient outcome. Also, the tumor and its microenvironment are characterized by spatial patterns, such as the arrangement of cells involved in the interaction of the immune system with the tumor, which cannot be adequately characterized by genomic approaches, yet is known to be highly relevant for prognosis. Nature selects for phenotype, not genotype (Gillies *et al.*, 2012). Even before the first human genome was fully sequenced, it became obvious that hopes and salvation expectations projected onto the emerging field of genomics would be challenging to fulfill.

The concept of the phene, coined a century ago, gained novel significance after being overshadowed by genetic approaches for years. An independent discipline named phenomics was postulated to help elucidate the physiological hierarchies leading from genetic traits to clinical endpoints (Schork, 1997). A decade later, it was understood that the systematic study of phenotypes in relevant biological contexts would be "critically important to provide traction for biomedical advances in the post-genomic era" (Bilder *et al.*, 2009). The toolbox of comprehensive phenotyping is ample and addresses complexity on different scales. From transcriptomics and epigenomics over proteomics to high-throughput assays of cell cultures, biological entities of increasing hierarchical complexity are assessed systematically (Houle *et al.*, 2010). However, these approaches are compromised by methodological challenges. Performing phenomics on a large scale is an intrinsically multidimensional problem, because the phenome is highly dynamic. It is influenced by a multitude of factors from post-translational modifications to high-level environmental stimuli. Approaches such as proteomics, capturing snapshots of multifactorial biological processes, are limited in their transferability and significance. In general, the gigantic molecular network that acts between the genome and the unfolding of tissue structures with all their properties is still not fully understood. Tissue phenomics provides a systematic way of knowledge discovery, thereby contributing to fill this knowledge gap.

1.2 Tissue Phenomics and Medicine 4.0

Tissue phenomics has also to be seen in the general trend of digitization and intelligent data processing. Digitization alone would not have such a big impact on society, science, and business. The combination of digitization with intelligent data processing and the availability of data networks with gigantic data streams changes our world. The most important keywords for this general critical endeavor are Industry 4.0, machine learning, automatic cognition, intelligent processing, cloud computing, data mining, and big data. In the medical domain, all those terms are also central, but additionally Medicine 4.0, digital medicine, digital pathology, and image analysis are themes that are specifically important. All those general and specific aspects and components are essential for tissue phenomics and will be discussed in this book.

The evolving digital word is changing our society in nearly all aspects. It is, therefore, an exciting question where this will lead us near-term and long-term. The principle changes in medicine will probably be not so different from the principle of general developments. The generation of knowledge and innovations is one example for this. In the past, machines could not do this; at best they could help humans to be creative and innovate. We live in a time where this is changing. Through automatic data acquisition, automatic data comprehension, data mining, and big data correlations, which represent potential discoveries of rules or laws, can be worked out by machines on their own. Automatically connecting those findings with what is known and documented in literature can lead to completely new insights. Through tissue phenomics the speed of creating medical knowledge and insights will be considerably enhanced. Clinical decision support systems (DSSs) have been in discussion for decades. Today it is obvious that very soon they will become very important and not so far from now they will contribute such crucial information that they will be indispensable for each and every therapy. This book illustrates why tissue phenomics will be a central part of this.

Digitization and intelligent data processing are playing a more and more dominant role for present and future businesses while also increasingly influencing daily life. Industry 4.0 is a prominent example of this development. Digitization of information is advancing vigorously and, combined with its intelligent processing and its massive collection and exchange with ever higher speed, is about to change our world probably more than anything else. We are moving toward a Society 4.0, and Medicine 4.0 will be an important part of it with tissue phenomics being embedded therein. Tissue phenomics, which in principle also could be called Pathology 4.0, is in the process of developing a power that goes far beyond conventional pathology. It is about the most important elements of living organisms, the cells, their states, and their interactions. It is already important today, but might even become the most important cornerstone of future healthcare.

Cells are the basic elements that form living organisms. Besides fulfilling all kinds of functions, they form an extremely sophisticated interaction network with many different types of players being involved. Therefore, it appears very reasonable to speak about a *social network of cells*. The value of tissue phenomics results exactly from getting comprehensive access to this interaction network. Tissue phenomics helps to understand it by extracting information and knowledge from it and by characterizing individual organisms. This is of general importance but is also crucial for the treatment of individual patients. Understanding in general the extremely complex interactions of cells by considering the different cell types and their different states is an essential part of biological and medical science. On the other hand, as this social network of cells differs from person to person, it needs to be characterized also for individual patients to enable their appropriate treatment.

Already today the importance of automatic rich quantification and comprehension of this social network of cells is clearly recognized by many scientists and clinicians, and its value is understood. Nevertheless, though the field is quite young and dynamic, it is worthwhile to consider also its potential future developments. In the near-term, it is obvious that automatic quantification and comprehension will become a standard procedure in science as well as in the clinic. The level of sophistication of the enabling technologies, of image analysis and data mining, is already very high and also improving fast. Furthermore, the process of including and integrating very different types of data besides the one derived from tissue slides has already started. In the long-term, the development of tissue phenomics will, in principle, be not so different from Medicine 4.0 and Industry 4.0. As these are very fundamental changes, the question is how far we really can foresee how those fields will develop.

1.3 Phenes in Cancer Immunology

Let us first concentrate on the present and near-term situation and on histology. An important type of cell–cell interaction is the interaction of the immune system, the immune cell community, with cells that are abnormal and cause a disease. Cancer represents such a case. Cancer cells interact with immune cells and have an influence on the state of immune cells and how they interact with each other and in return with the cancer cells. Knowing that the immune system, in principle, should be able to fight cancer cells, it became clear that the immune system in cases of cancer occurrence is not properly working anymore. Supporting the immune system in this fight against cancer is a relatively novel concept, but in recent years turned out to be extremely successful. This type of therapy is called immunotherapy and perhaps has even the potential to cure cancer. When studying the interaction of the immune cell community with the cancer cells, it became clear that this interaction is complex and that it differs from patient to patient. It is obvious that this is not only true for cancer but also for other types of diseases. There are many different diseases where the immune system plays an important role. Many diseases are simply caused by autoimmune reactions, and in cases of organ transplantation, getting the immune system under control is the biggest challenge. With the interaction network of cells, the *social network of cells*, being the essential part of any living organisms and that diseases are a result of this network being perturbed, it is not surprising that characterizing this social network of cells is the most important task for determining the right treatment of patients. Additionally, it has been shown that the treatment-decisive molecular heterogeneity of cancer cells in a single patient may be attributed to environmental selection forces, imposed by the network of cells constituting the tumor's immune contexture (Lloyd *et al.*, 2016). Today, the treatment decision is mainly achieved by histological investigations of individual cell populations, but in future, the study of networks of cells empowered by tissue phenomics will bring these decisions onto a new level.

1.4 Future of Tissue Phenomics

In the future, the interaction network of cells in tissue will also be characterized by non-microscopic techniques. Histopathology deals with investigating tissue slides under the microscope, where individual cells and their interactions can be studied. Tissue phenomics builds upon this. In most cases, single interaction events or cell states are not so relevant, but the statistical values of many interactions and states may be quite relevant. Therefore, if cell states and cell–cell interactions could be marked and visualized with a lower resolution by radiological investigations, so that statistical evaluations become possible for relevant regions like the tumor microenvironment, this could strongly complement microscopic histopathology. Through tissue phenomics, both types of data could be analyzed and the results could be combined. This is just one example for how tissue phenomics will evolve. Certainly genetic data are very helpful for evaluating states and interactions of cells. Here the same problem of low spatial resolution occurs, and again through combination with microscopic results, data become more meaningful. The trend is anyhow toward microlaser dissection where at least some local spatial data can be collected. In general, one could say the combination of results from very different data sources is what cognitive digitization will aim to achieve. For tissue phenomics, this means combining rich spatial information from high-resolution microscopic images with genomic and proteomic data with low spatial resolution, but rich depth, and with radiomics data providing whole organ contexts and longitudinal information. The integration dramatically increases the dimensionality of the information space with novel challenges and opportunities for big data analytics.

1.5 About This Book

The book first guides through the tissue phenomics workflow, as illustrated in Fig. 1.1, and second sets tissue phenomics in the context of current and future clinical applications.

Figure 1.1 Tissue phenomics provides a comprehensive workflow aiming at the discovery of the most accurate tissue-based decision support algorithm by close integration of assay development, image analysis and bioinformatics, and optimization feedback loops.

In Chapters 2 to 4, our authors describe in detail various approaches on how to convert the wealth of tissue slide pixel data into mineable knowledge (datafication). The journey begins with knowledge-based methods, in particular, the Cognition Network Technology (CNT) developed by Gerd Binnig and his team. Although knowledge eventually drives every analysis process, it became obvious in the last years that we have limited ability to translate the human (pathologists') knowledge about visual perception and recognition of objects in images to computer language. Therefore, data-driven approaches such as deep learning became increasingly important, as discussed in Chapter 4. The editors truly believe that the combination of knowledge-based and data-driven methods will eventually provide the highest impact on the utilization of image data for optimal treatment decisions for patients. Subsequently to the datafication of images, bioinformatics plays a crucial role in integrating other data sources such as genomics, radiomics, and patient-related information, and in generating prognostic and predictive models for disease progression. As discussed in Chapters 5 and 6, these models may classify patients in distinct groups such as those responding to a given therapy, or those with longer-than-average disease-related survival time. Since tissue phenomics provides a huge set of potential prognostic features (phenes), both chapters focus on robust feature selection methods by advanced Monte Carlo cross-validation algorithms.

In Chapter 7, we discuss multiple application examples of tissue phenomics in academic and commercial settings. In particular, Table 7.1 shows the tremendous impact of that approach to advances in biomedical sciences. Building on the successes in research, Chapters 8 and 9 discuss applications in clinical environments and provide a flavor of where our journey aims to. Translating tissue phenomics into the clinics is a demanding challenge, considering all the regulatory requirements and the novelty of the approach.

Finally, Chapter 10 looks into the future, where tissue datafication and subsequent patient profiling is part of every routine examination, with the goal to best match patients with the most successful therapy, as predicted by a tissue phene. This concept goes far beyond companion diagnostics, since it bridges from multivariate diagnostics to multiple therapy options offered by various pharmaceutical companies.

References

Bilder, R. M., Sabb, F. W., Cannon, T. D., London, E. D., Jentsch, J. D., Parker, D. S., Poldrack, R. A., Evans, C., and Freimer, N. B. (2009). Phenomics: The systematic study of phenotypes on a genome-wide scale. *Neuroscience*, **164**, 30–42.

Gillies, R. J., Verduzco, D., and Gatenby, R. A. (2012). Evolutionary dynamics of carcinogenesis and why targeted therapy does not work. *Nat. Rev. Cancer*, **12**, 487–493.

Houle, D., Govindaraju, D. R., and Omholt, S. (2010). Phenomics: The next challenge. *Nat. Rev. Genet.*, **11**, 855–866.

Lloyd, M. C., Cunningham, J. J., Bui, M. M., Gillies, R. J., Brown, J. S., and Gatenby, R. A. (2016). Darwinian dynamics of intratumoral heterogeneity: Not solely random mutations but also variable environmental selection forces. *Cancer Res.*, **76**, 3136–3144.

Schork, N. J. (1997). Genetics of complex disease. *Am. J. Respir. Crit. Care Med.*, **156**, S103–S109.

Chapter 2

Image Analysis for Tissue Phenomics

Johannes Zimmermann,[a] Keith E. Steele,[b] Brian Laffin,[a] René Korn,[a] Jan Lesniak,[a] Tobias Wiestler,[a] and Martin Baatz[a]

[a]Definiens AG, Bernhard-Wicki-Strasse 5, 80636 Munich, Germany
[b]MedImmune LLC, Gaithersburg, MD, USA
jzimmermann@definiens.com

In the process of tissue phenomics, image analysis is the sub-process of extracting histomorphological data from virtual slides, which are subsequently funneled into the data mining engine. Obviously, this procedure (also referred to as datafication) marks a crucial step in the chain of events. It is, therefore, discussed at length in this chapter.

By extracting and quantifying the immense number of biological entities present on multiple structural levels in a histological section, the foundation is laid for the identification of patterns and arrangements that are often difficult to quantify or simply not amenable to human assessment. The chosen image analysis approach has to cater to the complexity of the biological question being addressed. This can be a considerable challenge for certain disease settings such as the tumor microenvironment, which is composed of multiple cellular and molecular players.

Tissue Phenomics: Profiling Cancer Patients for Treatment Decisions
Edited by Gerd Binnig, Ralf Huss, and Günter Schmidt
Copyright © 2018 Pan Stanford Publishing Pte. Ltd.
ISBN 978-981-4774-88-8 (Hardcover), 978-1-351-13427-9 (eBook)
www.panstanford.com

10 | *Image Analysis for Tissue Phenomics*

However, only if the image analysis yields high-quality results and numbers are solid, downstream discoveries can be significant and robust. Objective knowledge about the performance of the employed image analysis solution with respect to a ground truth is thus critical. So too are a range of pre-analytical parameters that can facilitate a successful image analysis, as we also discuss briefly.

2.1 Introduction

The notion of personalized or precision medicine, grouping patients according to diagnostic test results in order to identify the most appropriate course of therapy for each patient, has been building over the past few decades due to the increasing availability of both therapeutic options and related diagnostic assays. This is especially true in the field of oncology. The value of a personalized medicine approach is exemplified in two recent US Food and Drug Administration (FDA) approvals for companion cancer diagnostics. One is the approval of Pembrolizumab in microsatellite instability high (MSI-H) tumors. Second is the approval of the Oncomine Dx Target Test for the selection of non-small cell lung cancer (NSCLC) patients encompassing three separate genomics-targeted therapeutic regimens. These also represent departures from standard cancer diagnostics in important ways. In the case of MSI-H testing, multiple cancer types can exhibit this phenotype; therefore, standard cancer diagnostic criteria essentially become overridden. The Oncomine Dx approach represents a multiparametric approach to diagnostic testing, in contrast to the standard one drug/one companion test model. It seems reasonable to envision a number of additional CDx tests based on multiparametric measures that address one or more tumor indications and assess multiple therapeutic possibilities in tandem. Importantly, this approach essentially employs big data techniques to wrap as many drug targets as possible into a single decision-making matrix. For health care providers facing an ever-growing list of available targeted therapies, immunomodulatory molecules, and combination therapies, this broader approach may be the best way to ensure that each patient ends up paired with the best therapeutic option for their particular disease.

In this context, genomic or gene expression analyses represent a large proportion of cancer-related diagnostic tests and are amenable to a multiparameter approach. However, there are practical limitations in using these analyses fully for standard of care tumor samples, classically represented by formalin-fixed, paraffin-embedded (FFPE) specimens intended for diagnostic histopathological analysis. Immunohistochemistry (IHC) assays also make up a substantial proportion of available cancer tests and readily make use of FFPE tumors. Multiple IHC assays are approved as either companion or complementary CDX tests. However, IHC tests currently are often made quantifiable in a simplified (monoparametric) way in order to optimize reproducibility among the pathologists who score them. Image analysis, by taking automated IHC analysis and tissue context into a more quantitative realm, can produce richer datasets for data mining and conceivably to identify novel diagnostic readouts. The greater complexity allowable by image analysis also places greater demands on data mining, essentially requiring a big data approach more along the lines of gene-based assays. In the future, personal medicine should not be constrained to a single platform, but rather could blend genomics, IHC, and other assay formats into a unified matrix.

Tissue phenomics is the term we apply to encompass the rich image and data analysis components of suitably designed multiparametric tests. On the front end, tissue phenomics must be supported by a digital pathology platform capable of generating the highest quality image analysis data possible. On the back end, it is made powerful by integrating image analysis data with all other available information, to include the results of any molecular testing. Feeding the resulting large datasets into big data and machine learning workflows has the potential to uncover signatures associated with response or resistance to a broad range of drugs and eventually could coalesce into tests, which could encompass the full range of therapeutic options for a given tumor type.

Beyond direct clinical applications, the tissue phenomics approach is also capable of yielding powerful insights into the pathogenesis of many diseases, as well as to enable early drug discovery efforts or support translational biomarker research. Widespread adoption of whole-slide scanning and a growing number of user-friendly image analysis software options have driven a sharp increase in published image analysis studies, which in turn has expanded awareness of the unique insights that can be gained from image analysis. Biomarker quantitation and spatial analyses outside the abilities of a human observer can be carried forward from pre-clinical research into the translational and clinical arenas, enabling biomarker strategies to take shape earlier and more efficiently. A number of tissue phenomics studies have demonstrated this ability to generate unique biomarker insight, improving existing manual assay interpretation or even succeeding where it failed entirely.

In this chapter, we discuss the image analysis component of tissue phenomics studies, together with the pre-analytical variables that are critical for the successful execution of these studies. A thorough understanding of the journey from tissue to data is essential for the proper implementation of tissue phenomics, especially in the context of precision medicine.

2.2 Experimental Design for Image Analysis Studies

Using image analysis, tissue phenomics is well suited to address complex biological questions. For instance, it can utilize highly multiplexed tissue techniques such as immunofluorescence, co-registered serial IHC, or imaging mass cytometry together with image analysis in a variety of ways. These kinds of multifaceted approaches can add great value to the data needed to optimize tissue phenomics. Complexity, however, also increases the potential to introduce noise into a system. Sources of noise must, therefore, be minimized. In short, image analysis is best when the experimental design and its execution maximize the extraction of all potentially quality data and at the same time minimize the effect of potential sources of problematic data. However, it is not necessarily obvious and straightforward what the experimental design and execution plan should be. In that regard, some pre-testing, validation, and, if needed, a revision may be required before settling on the final elements of the experimental design. We briefly discuss a few points to illustrate how design elements can influence the success of the image analysis study, as follows.

First, the general approach must be aligned with the needed data output. For example, a proximity analysis between two given cell types might be appropriately done by co-registering serial sections of chromogenic single stains. In contrast, measuring the dual expression of two markers in individual cells requires instead a dual stain. If the cellular compartments expressing the two markers are distinct (*e.g.*, one membrane and one nuclear) and can readily be discriminated by image analysis, then a chromogenic dual stain may be appropriate, whereas overlapping expression (two membrane markers) likely requires immunofluorescence. Second, image analysis can only reveal information present in the digitized tissue image. Therefore, all tissue and staining features needed to support the planned image analysis, including counterstaining for nucleus detection and incorporation of appropriate tissue masks (*e.g.*, cytokeratin for tumor cells), must be planned and validated upfront. Any needed information for which the image analysis cannot be automated might need to be addressed by other means. For example, manual annotations performed by a pathologist can readily partition data into relevant tissue regions. Third, minimizing potential problems should also be planned for as much as possible. For instance, image analysis may serve to reduce or eliminate some pre-analytical variables, such as adequately filtering out autofluorescence in immunofluorescence-stained tissues. But it may not account for other staining variables, such as the ability to sufficiently discriminate the brown staining of many IHC markers from the endogenous pigment present in melanomas. The staining part of the plan must, therefore, use alternative chromogens. It is not possible to consider all possible circumstances that could affect the outcome of a study. However, thorough planning and some degree of pre-study testing of what works and does not for the intended image analysis will greatly improve the chances of success. In particular, an early specification of the desired data points and what image analysis measurements can deliver those endpoints facilitates this design phase. In the end, a successful experimental design for a complex image analysis project must balance the potential benefits of the data to be produced against the potential risks of an overly complex set of methodologies and the possibility of introducing excessive noise into the data. In addition, other practical factors for consideration include logistical matters such as the computing resources required, transferability of scoring solutions to other tissues, or open re-use based on generated hypothesis.

2.3 Input Data and Test Data

As described in the preceding section, image analysis cannot be leaned upon to undo issues caused by pre-analytical variables. Therefore, correct sample handling, processing, staining, and scanning are as important as experimental design in determining the quality and reliability of image analysis data. Limiting cold ischemia, appropriate fixation time, automated processing, skilled sample embedding and sectioning, and use of an autostainer with antigen retrieval capabilities together provide the consistency necessary to generate the large, reliable image sets necessary for leveraging the full quantitative power of image analysis. In the context of tissue phenomics, this consistency is especially crucial to ensure that data mining and machine learning activities are performed on features calculated from tissues that have not experienced conditions that could alter tissue morphology or immune reactivity. Assay conditions themselves must be absolutely consistent; therefore, staining on a regularly maintained autostainer with in-run control tissues is strongly advised.

Ideal staining intensity for image analysis studies is often lower than for manual assessments, as maximizing dynamic range is more critical than highly visible staining; image analysis is not subject to the detection limits of the human eye, and intense staining conditions designed for binary pathology decisions can obscure information from the high and low ends of the staining intensity spectrum. Minimization of background staining is another reason to forego very intense stains, which can make accurate segmentation of individual cells in darkly stained areas problematic. Hematoxylin can be just as important to optimize as the stain itself, as nuclear morphologies and hematoxylin intensity can be invaluable tools in classifying both biomarker-positive and biomarker-negative cells into different cell types, which can then be aggregated into regional definitions. While CNT-based tissue phenomics image analysis approaches are not tied to nuclear objects as the starting point of an analysis, many simpler tools rely upon them for cell detection, meaning in these cases also optimized counterstains will give the best possible results from image analysis experiments. Tissue information made inaccessible through poor histology cannot be recovered by any image analysis system. Scanning of tissues is the final step in making tissue ready for image analysis and should be performed by experienced personnel on a high-quality scanner. Scans should be in focus for the entire slide if possible, and at a minimum, the regions of interest must be consistent and analyzable.

Quality control (QC) inspection of all images and tissue should be performed at this point, to ensure that the starting material for the image analysis is suitable. Visual and histological artifacts are removed at this point by manual annotation and automated means, to give the image analysis solution a head start by limiting the amount of non-target tissue it must interpret. Larger-scale regions of interest (ROIs) are also commonly annotated by a pathologist, eliminating the challenging image analysis task of automated ROI detection. However, annotations and QC of slides and image analysis results are time-consuming and subjective and require access to pathology resources. Automated QC methods trained and tested against expertly annotated "ground truth" of cells and regions in tissues can provide a way around the time and resources required for manual processes and are key to larger image analysis studies and tissue phenomics. Just as the correlation of multiple pathologist scores is measured for histologic assays, image analysis rulesets can and should be compared against pathologist annotations to ensure that performance of solutions is accurate and precise. These comparisons are likely to be critical for the entry of image analysis solutions into regulated environments such as companion diagnostics, where an extremely high level of performance is necessary to ensure patient safety and accurate treatment decisions. The image analysis solution, in this case, becomes part of the assay, and the entire process from start to finish should be tested with equal or greater rigor as an IHC test with a manual interpretation.

2.4   Image Analysis

An image analysis methodology that gained considerable traction over the last years (from 33 publications in the realm of digital pathology during 2005–2010 to 133 in 2016 alone) is CNT. The CNT

16 *Image Analysis for Tissue Phenomics*

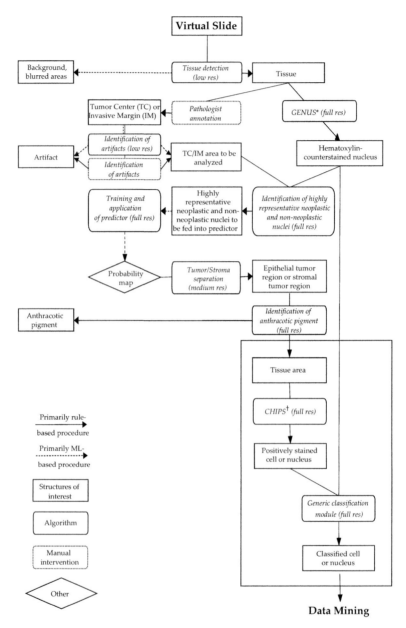

Figure 2.1 Schematic representation of a CNT-based image analysis procedure applied to a multiplexed virtual slide, combining the execution of both, rule-based and machine learning–based elements. (*Generic nucleus segmentation, †Chromogen-independent phenotype segmentation)

system perceives biological structures on all scales as objects, very much as a human observer would do, and allows a heuristic approach to complex biomedical questions (Binnig et al., 2002). Being object-oriented and context-driven, it is excellently suited for functional, morphological, and geographical phenotyping in histological sections (Baatz et al., 2009). When a virtual slide is analyzed, a fractal network of objects is generated, not only as final result, but already during an evolutionary process (Figs. 2.1 and 2.2) in which initial object primitives are semantically enriched until the final state of abstracted, correctly segmented, and classified objects of interest is reached (Fig. 2.3).

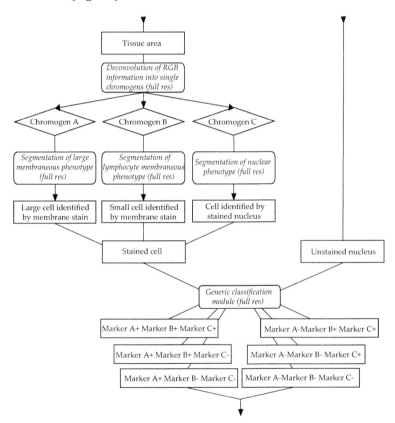

Figure 2.2 Detailed depiction of processes in boxed area of Fig. 2.1, highlighting the extraction of various cell species in the tumor.

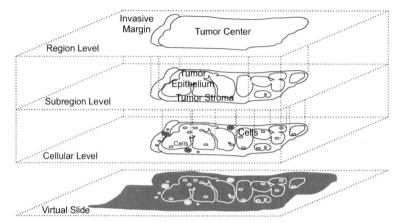

Figure 2.3 An example for a typical hierarchical representation of regions and cellular structures in Cognition Network Language (CNL). With this approach, different levels of granularity can be jointly represented in a common framework.

The procedure implies a constant oscillation between the spheres of (re-)segmentation and (re-)classification. Ephemeral structures might exist, whose only purpose is to serve as precursors for evolutionarily more refined, persistent structures, to enable measurements on specific regions or to efficiently handle large-scale structures on virtual slides, which might contain up to several gigapixels. The ongoing process is supported by a continuous evaluation of image objects along their spectral, morphological, and spatial–relational attributes. It allows generation of major mechanistic insights in (immuno)oncology, not only on virtual slides (Nagaraj et al., 2017) but also in all kinds of assays over the translational continuum (Estrella et al., 2013; Muñoz et al., 2017).

Pathological and biological knowledge from the expert is funneled in via the associated CNL (for a detailed treatise, see Chapter 3). CNL is a graphical language that allows an interactive scripting and offers immediate visual feedback on how the hierarchical object network is altered during the development process. The rule-based concept follows a highly semantic approach and facilitates a smooth and versatile development of image analysis protocols, rulesets, which can be enriched by a wide variety of machine learning approaches (see Chapter 4). The development environment for programming in CNL is Definiens Developer XD.

Image Analysis | 19

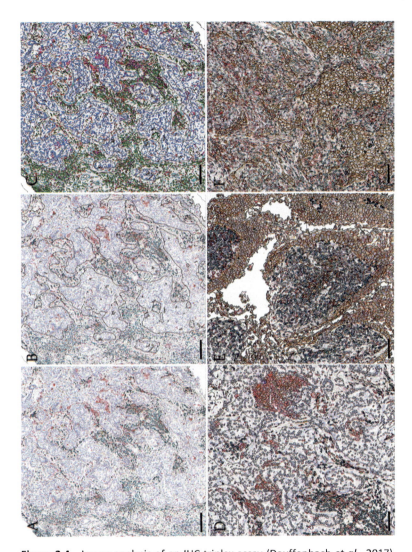

Figure 2.4 Image analysis of an IHC triplex assay (Dauffenbach *et al.*, 2017), including a macrophage marker (red chromogen), PD-L1 (DAB), and a marker for cytotoxic T cells (green chromogen). The image analysis follows the scheme exemplified in Fig. 2.1, resulting in an image object hierarchy as shown in Fig. 2.3. A–C: Image analysis of a case with low T-cell infiltration; scale bar corresponds to approximately 200 μm (A: Original image; B: Segregation of epithelial regions from stroma; C: Segmentation and classification of individual positively stained cells and counterstained nuclei, classification colors correspond to various cell species according to single or multiple marker positivities; D–F: Examples for the segmentation of individual cells in cases with varying marker expression; scale bar corresponds to approximately 100 μm).

In CNL, a straightforward grammatical concept is pursued, where the system itself—as a process-controlling instance—represents the *subject*. Processes, covering a panoply of granular image analysis algorithms, correspond to *predicates*. A domain concept allows addressing biological structures as *objects*, which are characterized by a class assignment and *adjective*-like descriptive features. These features are dynamically used in the course of the process as (transient) descriptors of biological structures but also play an essential role in the generation of final readouts, where they can be combined to yield advanced phenes, in some cases subvisual, context-derived patterns beyond human perception.

The principle of fractality is a leitmotif that pervades all aspects of the system (Klenk *et al.*, 2000) and is of crucial significance for rendering the multi-scalar nature of practically all biological phenomena. Biological ontologies become manifest in the image object hierarchy but can be as well represented using a hierarchical system of classes. Furthermore, features that are attributed to the biological structures and processes are organized hierarchically.

An archetypical application case for the image object hierarchy is the tumor microenvironment, where the tumor superstructure contains the tumor center and (sometimes) an invasive margin. The tumor center itself is made up of components such as tumor stroma and epithelial regions. Each of these entities again contains different species of cells, which can be grouped into substructures such as tertiary lymphoid structures. An individual cell might be composed of membraneous stretches, cytoplasmic portions, and a nucleus, which can hold a nucleolus or specifically textured chromatin microstructures. On a large-scale geographic level, the exact localization of intra-epithelial tumor-infiltrating lymphocytes (iTILs) can help to identify the exact character of a patient's immune response *sensu* (Teng *et al.*, 2015). On a cellular level, the hierarchical concept allows to understand where exactly within the individual cell the expression of a certain protein is taking place (Braun *et al.*, 2013).

The paradigm of the image object domain, implemented in the algorithmic framework, allows a dynamic navigation through the hierarchical network for local processing. The system can temporarily enable individual structures, like specific tumor regions, to become autonomous actors and execute particular actions on themselves or the structures they encompass, creating a dynamic ecosystem of objects. In the example of specific tumor regions, the cells they contain are consequently treated according to highly local rules. This enables targeted approaches for extracting objects of interest and can help, for example, in treating stromal and epithelial regions in different ways or to balance out staining gradients in histological samples.

The description of biological phenomena requires frequently the expression of expert knowledge in fuzzy terms (Zadeh, 1965). Membership functions help to address elaborate biological questions, such as differentiating principal modes of cancer invasion (Katz *et al.*, 2011) or identifying highly diversified cellular phenotypes in a robust manner (Azegrouz *et al.*, 2013) and can be used in the course of the datafication process in an explicit or procedural way.

Empirically, pathologists often approach the analysis of a histology slide by first gaining an understanding of the growth pattern of the particular tumor and evaluating the stain characteristics in multiple locations. Next, the assessment of the cells and regions is cast into percentages and/or scores, for example, for quantifying the KI67 proliferation of a tumor in breast cancer (Dowsett *et al.*, 2011).

Within the hierarchical framework described above, each object is able to query its location relative to neighboring objects and the hierarchy. Considering the example of TILs, a particular cellular object can be identified as TIL by examining its location within the hierarchy and by its spectral and morphological properties. Using a lymphocyte marker such as CD8 or by relying on morphologic criteria, objects on the cellular level can be categorized as lymphocytes or other cell species. Next, by querying the object hierarchy with respect to a particular object in question, the regional information can be used to derive a label; that is, if the detected cell is a subobject of the epithelial tumor region and indeed a lymphocyte, it can be labeled a TIL. Once a common representation of the regional and cellular objects is achieved, an abundance of scene statistics may easily be derived, such as cell densities stratified by type, proximity measures, or region-specific scores (Fig. 2.5).

As previously discussed, the CNT architecture encompasses a modeling of the tissue as a scene structure into hierarchical object levels comprising objects of varying granularity. Here, the knowledge about tissue preparation, disease characteristics, and staining quality is encoded during the development process (Baatz *et al.*, 2009). In contrast, recent advances in machine learning include deep learning architectures, which aim for automatically learning feature representations from training data, while relying on a predefined network architecture. In this manner, an automated learning of a hierarchical representation may be achieved; however, the network design and the availability of large datasets is crucial in order to arrive at robust and scalable solutions (Angermueller *et al.*, 2016).

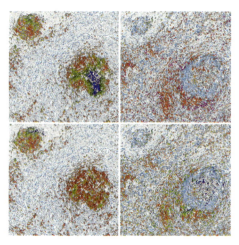

Figure 2.5 Proximity analyses on IHC-stained immune cell populations can be visualized as object-specific heatmaps. The degree of embeddedness of an individual positive cell with regard to another cell species on co-registered sections from the same block is represented as a color code, with values ranging from distant (blue) to proximal (red). Objects were identified using a CNL-based rule set. Left panel: Marker A positive cells and their color-coded proximities to two other markers (above and below). Right panel: Same analysis for lymphocyte marker B.

To this end, the modular and hierarchical approach of CNT allows to integrate machine learning approaches seamlessly with rule-based modeling approaches. In this case, machine learning approaches can be utilized to detect particular sub-entities in a tissue slide, for example, epithelial tumor (*e.g.*, the subregion level in Fig. 2.3) regions or cellular detections (*e.g.*, the cellular level in Fig. 2.3). Next, these individual entities can be merged using a hierarchical representation, allowing refinement and putting them into context. This approach has several advantages: First, the integrated representation of learnt primitives allows to fully exploit current state-of-the-art machine learning methods, enabling the best possible object detection or classification. Second, a hierarchical visual representation allows for structuring a tissue slide in accordance with the pathological assessment. An example of this would be categorizing cells analyzed via machine learning processes into those present in either invasive margin or tumor center, as in state-of-the-art immuno-oncology analyses.

This approach allows for the generation of analysis results in an explainable manner, that is, maintaining modeling up to the level required by the pathologists while allowing machine learning approaches to provide the best possible primitive generation. Using this scheme, the final representation of the hierarchy and objects allows for confirmation that an approach is biologically sound while generating statistics for subsequent data mining procedures.

The training of machine learning models requires large amounts of data, in general far larger than for the training of rule-based approaches. This can become a serious challenge in limited clinical datasets, where rarely does the training set cover all expected variations and expressions of the disease in question. The most straightforward approach to learn stable models is to acquire large amounts of annotated training data, learn a model and apply it. However, in practice, ground truth data might be difficult to achieve as, for example, manual cell annotations in sufficient quantity are hard to acquire or, for pairing outcome scores with tissue slides, sufficient numbers of slides are often not readily available. From this perspective, a hybrid strategy for analysis of tissue slides offers a substantial benefit, as a comparably simple rule-based approach can be utilized to train a subsequent, slide-specific machine learning model. Here, the rule-based approach generates training labels on the fly, that is, during analysis, while the classifier will allow to predict the remaining entities that were not captured by the rule-based image analysis. Similarly, locally concentrated tissue defects or alterations may be detected by simulating these from regular tissue, training a classifier on these and then finding similar regions. By using these hybrid techniques, the upfront need for (labeled) training data may be reduced, as domain knowledge is encoded into the image analysis. Expanding this principle, models pre-trained on other databases can be utilized to predict individual structures in a tissue slide, feeding into an analysis framework tailored to the image analysis problem in question. Once the ruleset is set up, a distributed client–server architecture allows its application in a production environment for batch processing of large numbers of virtual slides. The ruleset must be robust enough to cope with any type of biological and assay-related variability, noise, ambiguities, and level of detail of decisions from cell to cell, even making calls that are difficult for pathologists. It is self-evident that designing an image analysis solution requires an *a priori* knowledge of the expected variation in the training set. If the variability is adequately covered, it will perform robustly on large batches of sections. However, one day it might meet a *black swan* and fail, necessitating an adjustment of the algorithm to extend its applicability.

On a higher scale, robustness is of essential significance for broader tissue phenomics studies, like lateral tumor surveys over multiple cancer types: Readouts have to be comparable within and across assays and indications (Steele *et al.*, 2017). Therefore, analysis approaches should be as generalized and at the same time as modular and encapsulated as possible to facilitate transfer between solutions for various assays. They should also be as chromogen independent as possible to account for the advent of novel, multiplexed assays.

A number of analyses can be performed on digital images at a time well after the initial study design or early data generation and analysis. However, the ability to do such *post hoc* analyses may be limited by a number of factors, such as the rulesets used for image analysis, the scoring solutions developed, or tissue annotations. Decisions made upfront can either permit or limit the kinds of *post hoc* analyses that might be considered later. As a simple example of this notion, a study may intend to primarily assess glomerular changes in a disease model based on a robust automated detection of glomeruli. Since some glomerular diseases (*e.g.*, focal segmental glomerulosclerosis) may exhibit regional differences (juxtamedullary versus cortical glomeruli), the ability to later assess such regional differences may either be supported by prior planning (*e.g.*, the simple and efficient annotation of cortical and juxtaglomerular regions) or not. Such pre-planning decisions, however, must balance the potential benefit of preserving such later analytical decisions versus the effort required to do so.

A key notion regarding the ability of image analysis to produce the kinds of data needed to answer the biological questions under consideration is simply to consider which among available means can both efficiently and accurately produce such data. Automated software solutions, when they exist and can be easily applied, represent obvious choices for inclusion.

Fundamental to the ability of any image analysis solution to produce data that accurately define the histological changes present in a given disease model is the capacity to transfer the critical histopathological information present in the digital images of diseased tissue (or even normal tissue) into robust digital scoring solutions. In other words, the scientific expert must be able to initially transfer such histopathological information to the software developer in such a way that the developer can code the software to capture these changes. Importantly, this process must be pressure tested to ensure that what the image analysis solution *sees* is sufficiently aligned with what the scientific expert is seeing. Quite often this requires a back-and-forth process that refines the image analysis solution until there is a consensus that the working scoring solution accurately captures the key tissue changes. Inherent in this notion is that the system is eventually trained using tissues that appropriately reflect the changes expected in test tissues. This kind of iterative learning process may not be required when a machine learning process can instead be used to get to the appropriate scoring solution. It must still be kept in mind that the learning set of tissues or even the early set of test tissues may not fully represent all the histopathological variations representative of the disease condition of any particular population (*e.g.*, cancer of the lung). Therefore, some continued evolution of the initial working image analysis solution is often necessary that goes hand in hand with testing of the most recent algorithms. However, when the image analysis solution is finalized, it is also critical to validate that the data it produces accurately reflect the histopathological changes present in all of the test tissues, as we describe separately.

2.5 Quality Control Procedures

To achieve comparable readouts across assays and indications, all different sources of variability need to be understood and considered early in the process of developing the image analysis solution. Only then can a QC process have an impact on decisions made regarding the design of the image analysis solution. An adequate procedure should, therefore, consider both variability of input data as well as variability caused by the image analysis solution.

When investigating input data, it is highly recommended that a number of steps are performed by both a pathologist and an image analysis expert. These comprise identification of missing slides, visual identification of slides that do not belong to the same case, visual assessment of the preparation quality (sufficient preparation quality to visually identify cells and visual assessment of the staining quality), low-staining background, clear counterstain (*e.g.*, hematoxylin), visual assessment of the scanning quality, absence of scanning artifacts, and completeness of tissue scanned.

It is not uncommon that during the QC process of input data, some of the data need to be fixed, or recreated (*e.g.*, rescans of slides). Therefore, the topic of data management is also a very important one. It needs to be ensured that the correct versions of files are used and their history is traceable.

The other main category of QC is the assessment of the quality of image analysis results. This is required during development of the tissue phenomics image analysis solution as well as when assessing the quality of the readouts provided by it. To assess the quality, it is key that the task is well defined. Especially in the context of segmenting and classifying millions of cells, it needs to be understood what is good enough. Does the readout need to be continuous for the given task, or will a cut-off be applied on it to define two groups? In the case of the latter, the cut-off has a huge impact on the requirements of the image analysis solution as well as on the QC. The image analysis solution would have to be very sensitive around the cut-off, while far away from it sensitivity would be less critical. For example, if the readout is a percentage and the cut-off is around 20%, the test data should be enriched around this cut-off. On the one hand, the image analysis solution would need to distinguish above and below 20%. On the other hand, it would not matter if the image analysis solution would return 60% where 90% or 30% would have produced the same test result. This is important, as a better agreement at high cut-offs might come at the cost of having a poorer agreement at lower cut-offs.

When it comes to how to test the quality of the image analysis readouts, we can distinguish between visual QC by the expert (*e.g.*, a pathologist) and automated QC either based on ground truth or technical tests. Visual QC usually comes with relatively small initial costs as compared to creating automated test tools and acquiring the required ground truth for them. However, especially when doing QC during development, the same test might be applied multiple times. Furthermore, if the image analysis solution needs to be revised, automated tests can be easily applied to ensure the quality was not affected in a negative way. Additionally, visual tests are most often qualitative, or at most semi-quantitative, while automated tests can easily be set up in a way that allows quantification. To ensure comparability across assays and indications, the following must be demonstrated. First, it is possible to create a ground truth in a way that is comparable across assays and indications. To this end, the testing method needs to be clearly defined, trained, and tested for suitability. Second, the image analysis solution needs to be tested against this ground truth.

One example of an automated test is the Automated Classification Assessment (ACA). The ACA is intended to assess the quality of an image analysis result by quantifying the agreement between classified image objects and pathologist annotations. To this end, cell annotations are acquired by either one, two, or three independent pathologists using Definiens VeriTrova.

Pathologist-generated cell annotations are compared against image analysis results (classification) generated using Definiens Developer XD (see Section 2.4) via a process using Definiens VeriTrova, additional in-house software, and R (Lehnert, 2015; Stevenson, 2017; Neuwirth, 2014; R Development Core Team, 2016; Therneau, 2015; Therneau and Grambsch, 2000; Wickham, 2009). In case annotations are available from a single pathologist only, a single test is performed, assessing the agreement between this pathologists' annotations and the Definiens classification.

In case annotations are available from multiple pathologists, two different test scenarios are used. In the first scenario, quality is assessed for every combination of Definiens classification to pathologist annotation, alongside an inter-pathologist comparison.

In the second scenario, the classification is tested against a consensus of the pathologist annotations.

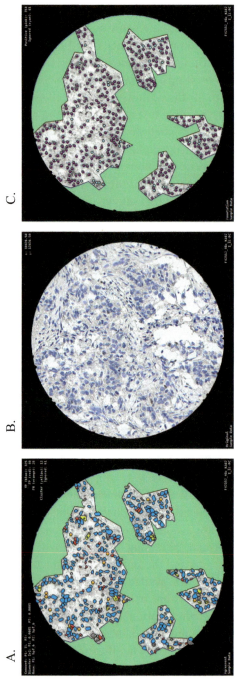

Figure 2.6(A–C) Illustrations showing (A) agreement (blue, TP; red, FP; orange, FN; yellow, cluster; black outline, ignored; green, excluded area), (B) original FOV, and (C) consolidated annotations (pink, positive; cyan, ignored; green, excluded area) for a given FOV.

Quality Control Procedures | 29

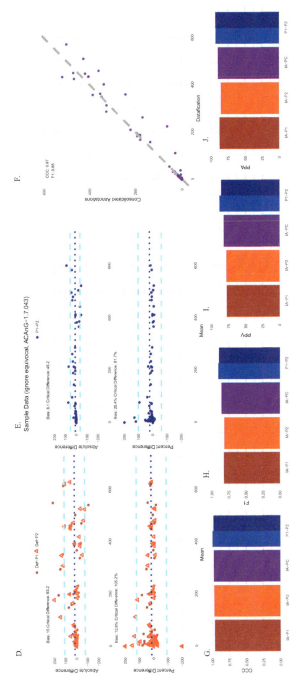

Figure 2.6(D–J) Result plot summarizing all derived measures of agreement of the ACA test. D: Bland–Altman plots of absolute and percent differences of counts per FOV for the agreement of counts per FOVs between classification and annotation(s); E: Bland–Altman plots of absolute and percent differences of counts per FOV for the agreement of counts per FOVs between annotation(s); F: Scatterplot of counts per FOV between classification and annotation or consolidation; G: Bar Plot of total CCC for the agreement of counts per FOVs between classification and consolidation, classification and annotations, and annotation(s); Bar Plots of total F1 score (H), PPV (I), and PPA (J) for the total count of TP, FP, and FN over all FOVs for the agreement between classification and consolidation, classification and annotations, and annotation(s).

Each test provides the following measures of agreement (see Fig. 2.6): For counts of target structures per FOV, Lin's correlation concordance coefficient (CCC, Lin, 1989) as well as Bland–Altman plots (Bland and Altman, 1986; Giavarina, 2015), and for the total agreement on a cell-by-cell basis F1 score (Powers, 2011), positive percent agreement (PPA) and positive predictive value (PPV, see Fig. 2.6).

To determine agreement with the ground truth in the context of comparability, the CCC should be used instead of the commonly used Spearman's rank or Pearson's correlation coefficients (SCC, Spearman, 1987; PCC, Pearson, 1895). While PCC and SCC are blind with respect to systematic over- or undercounting, CCC is not. Therefore, CCC is better suited to gain confidence in comparability of results.

To assess the strength of agreement, the criteria proposed by McBride (2005) can be used. To this end, the lower one-sided 95% confidence interval of the CCC is considered (CCC_lower). A CCC_lower greater 0.9 can be rated as almost perfect, greater 0.8 as substantial, greater 0.65 as moderate, and below 0.65 as poor agreement. Once data have passed through these QC measures before, during, and after development, it can be analyzed and interpreted with confidence.

2.6 Results and Output

The first step in analyzing results is itself an exercise in QC, as results should be reviewed at a high level as a *sanity check* to ensure that results have been calculated and exported as expected based on project planning and the QC process. Missing values, impossible values (*i.e.*, negative areas, H-scores over 300), and contradictory results (*i.e.*, no tumor cells detected in a region defined as a tumor) can and should be caught by manual and/or script-based review prior to thorough analysis. At this point, it can also be helpful to review again published literature relevant to the image analysis study to gain an understanding of what the expected results might be, and which observations are potentially novel, or entirely off-base. It can also be of great value to ensure that the headings of columns and rows in tabular data outputs are script-friendly, using underscores instead of spaces and avoiding +, –, and other operational symbols. Once the results have passed this final review hurdle, QC/QM reports should be prepared to accompany the data, so that any return to the data for additional analysis or to address questions regarding its generation can be completed successfully. Typically, sections assessing tissue and assay quality, slide scanning, image analysis results, and data exports are assembled under the guidance of a board-certified anatomical pathologist into a comprehensive report with example images. These reports are of critical importance in data mining and meta-analysis of previous image analysis studies, to ensure that the approaches, quality standards, calculations, terminology, and data formatting allow for cross-comparability with other datasets being funneled into the process.

2.7 Discussion and Summary

At this point, image analysis solutions based on CNT have reached a sufficient level of performance to enter into the data mining and meta-analysis workflows of tissue phenomics (see Chapter 5), itself an iterative self-improving loop similar to the cycle of segmentation and classification found in CNT. This process is discussed in more detail in the following chapters and has myriad applications, including determination of cut-points for patient stratification, identifying features associated with drug resistance and recurrence, and deep dives into immunological tumor landscapes. When this process is applied in the context of biological outcome data, its transformative power in biomarker analyses can be fully realized. With consistent and thoughtful crafting of such studies entering the tissue phenomics loop, far-reaching insights spanning multiple indications, biomarkers, and therapeutic modalities can be had.

Acknowledgments

We gratefully acknowledge Mosaic Laboratories LLC for providing the virtual slides shown in Fig. 2.4.

References

Angermueller, C., Pärnamaa, T., Parts, L., and Stegle, O. (2016). Deep learning for computational biology. *Mol. Syst. Biol.*, **12**, 878.

Azegrouz, H., Karemore, G., Torres, A., Alaíz, C. M., Gonzalez, A. M., Nevado, P., Salmerón, A., Pellinen, T., Del Pozo, M. A., Dorronsoro, J. R., and Montoya, M. C. (2013). Cell-based fuzzy metrics enhance high-content screening (HCS) assay robustness. *J. Biomol. Screen.*, **18**, 1270–1283.

Baatz, M., Zimmermann, J., and Blackmore, C. G. (2009). Automated analysis and detailed quantification of biomedical images using Definiens Cognition Network Technology. *Comb. Chem. High Throughput Screen.*, **12**, 908–916.

Binnig, G., Baatz, M., Klenk, J., and Schmidt, G. (2002). Will machines start to think like humans? Artificial versus natural Intelligence. *Europhys. News*, **33**, 44–47.

Bland, J. M. and Altman, D. G. (1986). Statistical methods for assessing agreement between two methods of clinical measurement. *Lancet Lond. Engl.*, **1**, 307–310.

Braun, M., Kirsten, R., Rupp, N. J., Moch, H., Fend, F., Wernert, N., Kristiansen, G., and Perner, S. (2013). Quantification of protein expression in cells and cellular subcompartments on immunohistochemical sections using a computer supported image analysis system. *Histol. Histopathol. Cell. Mol. Biol.*, **28**, 605–610.

Dauffenbach, L. M., Sia, G. C., Zheng, J., Jun, N., Olsen, E. P., and Kerfoot, C. A. (2017). Multiplex immunohistochemical staining of PD-L1, PD-1, CD3, CD4, CD8, CD68, FoxP3, and Ki-67 and image analysis of tumor and invasive margin in human FFPE NSCLC tissue. *Cancer Res.*, **77** (Supp 13), Abstract 1669.

Dowsett, M., Nielsen, T. O., A'Hern, R., Bartlett, J., Coombes, R. C., Cuzick, J., Ellis, M., Henry, N. L., Hugh, J. C., Lively, T., McShane, L., Paik, S., Penault-Llorca, F., Prudkin, L., Regan, M., Salter, J., Sotiriou, C., Smith, I. E., Viale, G., Zujewski, J. A., Hayes, D. F., and International Ki-67 in Breast Cancer Working Group. (2011). Assessment of Ki67 in breast cancer: Recommendations from the International Ki67 in Breast Cancer working group. *J. Natl. Cancer Inst.*, **103**, 1656–1664.

Estrella, V., Chen, T., Lloyd, M., Wojtkowiak, J., Cornnell, H. H., Ibrahim-Hashim, A., Bailey, K., Balagurunathan, Y., Rothberg, J. M., Sloane, B. F., Johnson, J., Gatenby, R. A., and Gillies, R. J. (2013). Acidity generated by the tumor microenvironment drives local invasion. *Cancer Res.*, **73**, 1524–1535.

Giavarina, D. (2015). Understanding Bland Altman analysis. *Biochem. Medica*, **25**, 141–151.

Katz, E., Verleyen, W., Blackmore, C. G., Edward, M., Smith, V. A., and Harrison, D. J. (2011). An analytical approach differentiates between individual and collective cancer invasion. *Anal. Cell. Pathol. Amst.*, **34**, 35–48.

Klenk, J., Binnig, G., and Schmidt, G. (2000). Handling complexity with self-organizing fractal semantic networks. *Emergence Complex. Organ.*, **2**, 151–162.

Lehnert, B. (2015). Blandaltmanleh: Plots (slightly extended) bland-altman plots. Available at: https://cran.r-project.org/web/packages/BlandAltmanLeh/index.html

Lin, L. I. (1989). A concordance correlation coefficient to evaluate reproducibility. *Biometrics*, **45**, 255–268.

McBride, G. B. (2005). A proposal for strength-of-agreement criteria for Lin's Concordance Correlation Coefficient. *NIWA Client Rep.* HAM2005-062.

Muñoz, S., Búa, S., Rodríguez-Acebes, S., Megías, D., Ortega, S., de Martino, A., and Méndez, J. (2017). In vivo DNA re-replication elicits lethal tissue dysplasias. *Cell Rep.*, **19**, 928–938.

Nagaraj, A. S., Lahtela, J., Hemmes, A., Pellinen, T., Blom, S., Devlin, J. R., Salmenkivi, K., Kallioniemi, O., Mäyränpää, M. I., Närhi, K., and Verschuren, E. W. (2017). Cell of origin links histotype spectrum to immune microenvironment diversity in non-small-cell lung cancer driven by mutant Kras and loss of Lkb1. *Cell Rep.*, **18**, 673–684.

Neuwirth, E. (2014). RColorBrewer: ColorBrewer Palettes. Available at: https://cran.r-project.org/web/packages/RColorBrewer/index.html

Pearson, K. (1895). Note on regression and inheritance in the case of two parents. *Proc. R. Soc. Lond.*, **58**, 240–242.

Powers, D. M. (2011). Evaluation: From precision, recall and F-measure to ROC, informedness, markedness and correlation. *J. Mach. Learn. Technol.*, **2**, 37–63.

R Development Core Team. (2016). *R: A Language and Environment for Statistical Computing* (Vienna, Austria: The R Foundation for Statistical Computing).

Spearman, C. (1987). The proof and measurement of association between two things. By C. Spearman, 1904. *Am. J. Psychol.*, **100**, 441–471.

Steele, K. E., Tan, T. H., Korn, R., Dacosta, K., Brown, C., Kuziora, M., Zimmermann, J., Laffin, B., Widmaier, M., Rognoni, L., *et al.* (2017). Measuring multiple parameters of tumor infiltrating CD8+ lymphocytes in human cancers by image analysis. *Sci. Transl. Med.*, In preparation.

Stevenson, M. (2017). epiR: Tools for the Analysis of Epidemiological Data. Available at: https://cran.r-project.org/web/packages/epiR/index.html

Teng, M. W. L., Ngiow, S. F., Ribas, A., and Smyth, M. J. (2015). Classifying cancers based on T-cell infiltration and PD-L1. *Cancer Res.*, **75**, 2139–2145.

Therneau, T. (2015). A package for survival analysis in s. Available at: https://cran.r-project.org/web/packages/survival/index.html

Therneau, T. M. and Grambsch, P. M. (2000). *Modeling Survival Data: Extending the Cox Model* (Springer).

Wickham, H. (2009). *ggplot2: Elegant Graphics for Data Analysis* (Springer).

Zadeh, L. A. (1965). Fuzzy sets. *Inf. Control*, **8**, 338–353.

Chapter 3

Context-Driven Image Analysis: Cognition Network Language

Gerd Binnig

Former CTO and founder, Definiens AG, Bernhard-Wicki-Strasse 5,
80636 Munich, Germany
binnigerd@gmail.com

The Cognition Network Language (CNL) has been developed to enable context-driven analysis, in particular for images. Context-driven image analysis is required when the content of images is very diverse and complex. Analyzing images for sophisticated tissue phenomics applications is indeed a complex task. In tissue phenomics, the complex *social networks of cells* and the interaction of different cell types reflect the state of a disease and need to be captured in a sophisticated way. For this task, context-driven image analysis has been developed. Its elementary concept is the stepwise build-up of knowledge about the image to be analyzed, which serves as an evolving foundation for an increasingly advanced analysis toward the end of the analysis. In this chapter, the reasoning behind the context-driven analysis and its most important basic principles are explained in the first parts. After this, a particular image analysis

---

Tissue Phenomics: Profiling Cancer Patients for Treatment Decisions
Edited by Gerd Binnig, Ralf Huss, and Günter Schmidt
Copyright © 2018 Pan Stanford Publishing Pte. Ltd.
ISBN 978-981-4774-88-8 (Hardcover), 978-1-351-13427-9 (eBook)
www.panstanford.com

problem is addressed as an example for demonstrating some of the principles in more detail.

3.1 Motivation and Reasoning

CNL is a computer language that had been developed by Definiens for automatic understanding and quantification of complex data structures (Schäpe *et al.*, 2003; Benz *et al.*, 2004; Baatz *et al.*, 2005; Athelogou *et al.*, 2007). The principles of CNL can be applied to all kinds of data structures. To be most efficient and easy to use, CNL is tailored to specific problems. At present, a fully deployed CNL developer is implemented for images and image data but is also already in an advanced state for data mining—specifically for data mining based on results from image analysis. The focus of this chapter, however, is on CNL for image analysis.

CNL was developed for complex real-world images. The vast majority of images created today in the real-world—in contrast to model systems or handpicked image data sets—are very complex. There are many domains where image analysis adds value, but most desired and valuable are meaningful analyses of medical images. To gain deeper insights into medical and biological processes, the discovery and quantification of relevant structures and patterns in large sets of images are most helpful, needless to mention that analysis of tissue slides is a prerequisite for tissue phenomics.

For a broad approach to medical images for the daily routine, it cannot be avoided that they are in general very complex. If a problem is too complex to be solved exactly by an equation, other ways of solving the problem have to be invented. The famous traveling salesman is one of those problems. In this setup, a salesman needs to visit many different locations and wants to know the shortest path for visiting all locations. For a large number of locations, the number of different paths is huge. Like for analysis of images, the traveling salesman represents a combinatorial problem, where the number of possibilities of how to combine elements into a solution is too large to be calculated in a reasonable time. Therefore, other approaches than calculating simply all combinations and then picking the best one have to be investigated. One solution for the traveling salesman problem is an evolutionary approach in the form of genetic algorithms.

In image analysis, the combinatorial problem results from the sheer endless number of possibilities of how to group pixels into objects. In most of the image analysis tasks, the goal is to detect and quantify objects. For this, the pixels in a digital image have to be grouped into objects, a procedure called segmentation. The term *segmentation* results from the reverse way of thinking about the problem: cutting the digital image into segments. The result is the same; pixels are grouped into segments that hopefully represent relevant regions or objects.

There are situations where segmentation is simple. For example, when all relevant objects in an image are bright and everything else is dark, or vice versa. A simple algorithm, a threshold on the brightness of the image, is sufficient to solve the problem. In medical images, this type of contrast enhancement is often achieved on purpose for an easier visual inspection or an easier digital analysis. In radiology, a contrast bolus is often used, and in pathology, genes or proteins are stained to highlight specific functions or objects. This simplifies the problem but solves it only in specific cases. *Real-world* images are still complex. Contrast enhancement is not perfect. There are artifacts where objects or regions receive contrast enhancement that was not intended to be highlighted. Furthermore, the enhancement is in many cases very heterogeneous with different objects touching each other. This leads to the difficulty to decide where an object ends and a new object starts. There is a long list of problems. Some of them will be discussed by examples in the technology section.

The answer of the CNL approach to the complexity problem is similar to the solution of the traveling salesman problem. The problem is addressed through an evolutionary stepwise procedure. Here a path has to be found as well, which is, however, more complex and abstract than just visiting fixed given locations. In CNL, the stepwise procedure is called context-driven analysis. The *locations visited* are analysis procedures that produce intermediate results. The principle is to move from analysis procedure to analysis procedure. Let us call these individual analysis steps *context steps*. In each context step, a result is achieved that makes it easier to address the next context step. A particular previous context step serves as the context for the following ones. It is natural that the first steps need to be the easier ones, as they cannot be built upon much context. The path to be found for a specific data analysis problem is from the simplest context steps over medium difficult steps to finally solving the complex problem. This represents a path-finding problem, and it is not obvious which path to take. For real complex problems, the major context steps might even consist themselves of sub-paths, resulting in a somewhat self-similar algorithmic structure of the final computer code. Furthermore, the path might establish in a form that goes beyond a simple sequence of context steps. There might be several context steps required to successfully perform a succeeding context step. Therefore, in many cases, the context steps are linked in the form of a network. This becomes perhaps more understandable in the case where several context steps produce (segment) different types of objects that are used as context for detecting and classifying even more complex object types in their neighborhood. This network character contributed the N to CNL. Solving this way complex image analysis problems is more than conventional image analysis and rather represents a cognition process similar to human cognition. In the time when the basics of CNL was developed, from about 1996 to 2006, it was not clear to scientists what the meaning of context possibly might be. It, however, was clearly envisioned that context is important and that the human mind makes use of it, which leads to much more intelligent results than those usually achieved by computers.

Some simple context-driven procedures were already implemented early in the history of image analysis. In particular, in the image analysis of radiological images, such as CT images of humans, it was obvious that the spatial relations between the organs are quite well-defined and that one can make use of it for image analysis. It is obvious that for detecting the liver in a CT scan, the detection of the lung beforehand makes the analysis simpler. The air in the lung produces an excellent contrast to the rest of the body and is (also because of its size) in most cases easy to locate. Once the lung is found, it is obvious where to expect the liver. This is a very simple example of context-driven analysis. There is, however, a difference between the scarce usage of context and the introduction of the new concept of context-driven analysis. In the latter case, a specific technology has to be developed to simplify the intense usage of context-driven processing. In case the *liver problem* is treated in the form of a context-driven approach in a deeper meaning, one might even rethink the path from the lung to the liver. As a result of certain diseases or of lung surgery, detecting the lung might in some cases be even more difficult than the detection of the liver. In CT scans, the spinal cord, however, is an object that can be detected most easily and reliably because of its simple shape and high contrast to the neighboring spinal bone. Detecting it represents the first step in defining the coordinate system for locating the liver. The next step might indeed be detecting the lung. With the spine as context, it is, however, easier to discover problems with the lung by analyzing relative location and symmetry. This shows that a broad context-driven approach can result in a more advanced strategy.

Analyzing tissue slides also requires context-driven processing. It is, however, more difficult to define the network of context steps than for radiological scans. This spatial relational network of objects (organs) in CT scans does not exist in such a well-defined form for cells or cell arrangements. The brightness of objects in the different color channels of those images is also only poorly defined in contrast to CT scans. Therefore, besides detecting neighboring context objects for the more difficult to find objects, other types of context steps are required as well. Some of them will, therefore, also be used for evaluating the color tone, brightness, or contrast of a particular image and of objects therein. Also, finding early hints for what kind of tissue is analyzed (*e.g.*, advanced tumor or healthy tissue) might help to apply the appropriate algorithms. In general, a context step is a processing step that produces results that are useful for the next context step. It is likely that many of those processing steps are not relevant in the final outcome of the analysis, but they might be useful as context-generating steps.

In a context-driven analysis, semantics plays an important role. "The liver is located underneath the lung" or "a cancer nucleus might be enlarged and might contain several nucleoli" are examples of this. Lately through machine learning and in particular through deep learning, impressive results could be achieved without using any semantics. CNL is, however, not intended to compete with state-of-the-art image analysis algorithms. CNL rather makes use of those techniques and integrates them. A final CNL program or CNL solution represents a network of context steps. Some of them might be driven by machine learning procedures, others by semantics. The overall network, however, for solving the complex problem is formulated through CNL.

Introducing context-driven analysis as a new principle that requires a specific technology to address the problem in a broad and generic sense. CNL is such a technology.

In Section 3.2, the relation between the state-of-the-art image analysis techniques and CNL will be investigated, and in Section 3.3 on technology, the most important components of context-driven analysis by CNL are described.

3.2 History and Philosophy

Traditionally, image analysis was developed mainly for model systems and simple tasks. Images produced, for example, in the industry with complex content could not be analyzed. When Definiens first offered to analyze images taken from tissue slides, we were told this is not possible and only humans can analyze those images. For somewhat more complex tasks that go beyond just counting, for example, brown cells or nuclei, semi-automatic approaches would have to be taken. We just had developed CNL at that time and could indeed solve more complex problems (Baatz *et al.*, 2008, 2009). The technology still had some problems at that time (in particular, processing speed was not great), but most principles of context-driven analysis were already implemented. As mentioned earlier, semantics plays an important role in CNL. Semantics is the way of humans to describe the world. It is important for communicating knowledge. Artificial neural networks (NNs) cannot communicate what they have learned to other NNs. What has been learned by the NN is encoded in a large number of weights in the connections between the neurons. It is by no means obvious how to give a meaningful description of the difference between two NNs that have been trained differently on the same problem. This means they cannot learn from each other, at least not yet.

For people, this is different. People with their natural neural networks can indeed communicate and learn from each other. To a large extent we communicate and learn through semantics. However, not everything we know, we can communicate verbally. Sometimes we recognize something, but we do not know how we achieved it. It might take a long time to translate this knowledge into semantics. In school, semantics plays a dominant role. For a baby, however, semantics does not really exist but it still learns with a high speed. It is probably due to the same aspects through which an NN gets trained, that is, by examples and by repetitions. Interestingly, it does not require labeling, tagging, or *annotations* like most of the trainings of NN are set up. It learns how to distinguish mother from father although the mother might wear clothes that are somewhat similar to the ones the father was wearing a day earlier, and this works without tagging, that is, without the baby knowing about the cloth facts. It learns just by itself what the relevant features of an important person are without making use of tagging or labeling parts of the person or the person as a whole. Tagging might, however, occur through context. Seeing the same face might be correlated with breastfeeding, a similar voice, type of body contact, or smell.

This means learning or training does not necessarily require labeling or semantics. Today most NN procedures indeed use labeling but do not use semantics at all. The future of automatic data understanding lies probably in combining NNs with semantic processing and by making use of labeling as well as correlations and cluster analysis. This way a system could possibly be trained in a less labor-intensive fashion on complex problems than today. The different aspects have their individual pros and cons. Semantics is the way to train a system with what is known by humans. If something is already known, there is no need for complex training procedures. You tell the system similar to how you would tell another person. The combination of labeling or annotating with training of NNs represents a mix of using human knowledge combined with machine learning techniques. This is also a way how humans often communicate with each other, for example, by pointing at something and connecting a name to it. At Definiens, we have collected experiences over many years in transforming expert knowledge into the computer language CNL. This means translating human semantics into machine semantics. From this experience, we know that experts often do not know why they recognize something, but they definitely know that they indeed have recognized a relevant object. In this case, they explain what they see, categorize the findings, and simply point at the related regions or objects. In such a case where knowledge cannot be formulated semantically, annotating images and training the machine might be a good option.

In case of incomplete knowledge about a cognition process, people often use a mixture of semantics and pointing (annotation). It might be worthwhile to find out how to implement something like that in software. It is to expect that the increasingly advanced integration of machine learning into CNL will contribute the most to the progresses of CNL in the next years. In particular, the problem of how to find the optimum path for context-driven analysis through machine learning needs to be addressed. When dealing with the theme cognition, there will never be a final algorithmic solution, but rather always be room for improvements. This has to do with the complexity of the problems.

CNL has, however, been developed exactly for this task of

- Combining and integrating existing methods into one technology
- Providing ease of use techniques for transferring knowledge into computer programs
- Enable context-driven analysis

These are the basic elements that are required for solving complex tasks. CNL is the computer language for approaching the complexity problem. The original thought for developing CNL as a context-driven processing technique was that complex tasks cannot be solved in one step with one algorithm even if it is as sophisticated as it can be. Life, for example, developed into something extremely complex, in a stepwise fashion. There is an alternation of modification (mutation, variation) and learning experience (natural selection). There is a similar process for industrial products. Products are improved on the basis of what experiences have been made with previous products. Nobody could have designed a car of today a hundred years ago, not even by the deepest ingenious thinking. If there are too many possibilities, one cannot think about all of them in a reasonable time. Combinatorial problems cannot be solved within one step. The question is: What are the most relevant mechanisms for a stepwise evolutionary, context-driven approach?

CNL, as a technology, developed stepwise as well and is still in development as explained before. The first step was to implement object-based image analysis and a class hierarchy where explicit knowledge can be formulated in an easy way that is not too different from human semantics. Semantics is usually formulated based on objects, their properties, and their relations. Some of those relations are hierarchical relations like *is part of*. Other relations are neighborhood relations like *is neighbor of*. Those relations and the properties of objects have to be reflected in the architecture of the processing tool. It was implemented in the form of a hierarchical network of objects that can be created from the pixels and in the form of a class hierarchy, where properties of objects can be described by fuzzy logic in a natural form and hierarchical and neighborhood relations can be formulated as well. At the end of this chapter, in the acknowledgments, the key people in the research and development team are mentioned with their specific contributions.

3.3 Technology

In this section, CNL is described in more detail and the most important elements will be explained, sometimes also by example. First, object-based processing is described, as it perhaps represents the most important element in CNL. It is a prerequisite for context-driven processing. Relevant objects in an image are embedded in a hierarchical relational network of other objects, which represents their context. The essentials of these networks can be formulated through semantics. Without the concept of objects, semantic knowledge does not exist. Semantics represents condensed knowledge about the world. It constantly evolves. Once it is well established, it had been matured over at least some years. The expert knowledge, for example, the knowledge of a pathologist is not knowledge about image analysis but rather about the subject of what is imaged. It has to be turned into image analysis knowledge. One way of doing this is by transforming semantics of the expert into analysis-software-semantics. Here objects and their class description come into play.

There are two types of knowledge that can be formulated in CNL: knowledge about the processing path and knowledge about the objects that are intended to detect. The first can be formulated in the process hierarchy and the latter in the class hierarchy. In both hierarchies, the required object network for context-driven analysis is indeed reflected in how CNL is structured.

3.3.1 Class Hierarchy

In the class hierarchy, classes of objects can be defined through their object properties and context properties. There is a long list of predefined features that are available for their use in the class descriptions. Complex class descriptions can be formulated, if necessary. With logical expressions like *and*, *or*, and *weighted mean* values of fuzzy logic features, sophisticated decision trees can be formulated. Through fuzzy logic, the probability for a support of a class membership can be edited as a class membership probability that depends on the value of the feature. In the object hierarchy, which forms when the processes of the process hierarchy are executed, a hierarchical network is automatically created. This way all kinds of statistical context features from the network neighborhood become directly available. Here, by neighborhood a generalized, hierarchical neighborhood is meant, where a sub-object or super-object of an object is also viewed as a (hierarchical) neighbor. Additionally, desired types of networks can be formed through the creation of links. Special links between any kind of objects can be established through processes making customized context features also available.

There are two hierarchies established in CNL: a group hierarchy and an inheritance hierarchy. In the group hierarchy, classes can be grouped into a super-class and super-classes into super-super-classes, and so on. In the inheritance hierarchy, families of classes are defined where family member classes are similar to super-classes and their properties (class descriptions) are inherited from the super-class.

3.3.2 Process Hierarchy and Context Navigation

3.3.2.1 Processes and the domain concept

In the process hierarchy, specific processes can be formulated that define what kind of operation should be applied to the image or to the related data structure (*e.g.*, the objects). Those processes can be picked from a long list of predefined processes and can be parameterized. If the operation aims at image objects, subgroups of objects can be selected for the process. The definition of the object subset to be treated is called the object domain. By defining the class of the objects, a certain region within the image and certain conditions related to object features, a list of objects is selected among all available objects. (It is also possible to select a *map* where to find the objects, which will be explained later in this text.) In the same process template, besides the domain it can also be decided what kind of process with what kind of parameterization should be applied to the selected objects. By adding one process after the other, a long processing story can be told. When executed, one context step after the other is performed and finally the objects of interest are detected and quantified. One might ask the question why is this called process hierarchy and not simply a sequence of processes? One always could group the processes hierarchically to make the whole sequence look clearer and nicer and make it easier to understand. In CNL, the hierarchy is, however, indeed real and meaningful. This is related to the context-driven image analysis strategy. For this context, navigation is implemented. In general, navigation through the hierarchical neighborhood or, in other words, through the hierarchical network is essential for context-driven analysis. This possibility of context navigation represents probably the most unusual and, at the same time, the most valuable feature in CNL.

3.3.2.2 Context navigation

There is the option to formulate hierarchical processes where a navigation through the hierarchical object network gets feasible. It starts by defining the type of objects where the navigation should begin, that is, by defining a domain. Now through a sub-process, it can be defined where to navigate to next from each of those objects in the domain. There is the option to navigate one or more hierarchical object levels up or down or to direct neighbors on the same level or to neighbors below a certain distance away from the domain object. This is not necessarily the end of navigation. With a further sub-process of the sub-process, the navigation can continue, and so on. In each navigation step, class memberships and conditions can be formulated for the objects to be navigated to, just like for the domain of the starting objects. In principle from each starting object, every other object in the scene can be reached. In each navigation step, there is the option to apply a process. For instance, objects can be reclassified or statistical information can be collected from the generalized neighborhood. This can be used, for example, for context-driven classification. Classifying immune cells is one example. One could start with a seed object, which is an immune cell with an excellent classification. Some features of the starting object can be written into a variable and when navigating to a *perhaps-immune-cell* in the neighborhood (*perhaps* because of a somewhat distorted shape), its features can be compared with those of the starting object. If these features beside the shape are very similar (this might be the domain condition), then the *perhaps-immune-cell* might be reclassified into an immune cell. The reverse would also work: Starting with an uncertain immune cell and navigating into the neighborhood, several clear-cut immune cells might be found and a statistical evaluation about some of their features might help to reclassify the uncertain one into a clear-cut immune cell. This is actually a strategy pathologists apply when they classify immune cells manually; they use this type of context. With the help of domain navigation, context information can be used constructively.

This is not only true for classification. In particular, for re-segmentation context, navigation can be helpful. There is always the possibility that an error occurred in the segmentation. There might be a little artifact where the border of a segmentation got trapped. Searching for alternative segmentations might solve the problem. This, however, is expensive with respect to processing time. Context might tell that in certain regions this effort might be worthwhile.

3.3.2.3 Object-based processing

When Definiens first brought its technology to the market, the scientists working on image analysis of satellite images (Benz *et al.*, 2004) introduced the term *OBIA* (object-based image analysis). At that time, a community formed around this term and nearly all the people in the community worked with *eCognition*, the early implementation of Definiens' context-driven image analysis method. Image analysis of satellite images is, in most cases, also a very complex task and context-driven analysis is required.

In general, *object-based* does not only mean that objects are created and classified. There is sometimes a confusion associated with this term. Detecting objects per se is nothing special. In most image analysis technologies, objects are segmented and classified in the end. In context-driven analysis, however, also intermediate objects, or objects that might not be of direct interest, are detected and used for the detection of the actual objects of interest.

Furthermore, the segmentation of the images can, besides pixel processing, be performed on the object level, which means that the objects generated in the first step can be processed further. For example, instead of grouping pixels into objects, objects can be grouped into objects. A very simple example of that kind of object-based processing that differs clearly from pixel processing is shown in Fig. 3.4 and described later in the example section. Here a nucleus is not yet detected, but rather intermediate objects that can possibly be combined into the object of interest, the final nucleus. This can be a very fast process, as there are much less objects than pixels. Here the concept of a *Parent Process Objects* (PPO) comes into play, which does not exist for classes defined on the pixel level. For a model system with a single seed object being analyzed, this makes no difference. Reality looks different and seed objects of the same class might be densely packed and the candidate object that fits nicely into the indentation of the seed object might also touch another seed object. For the decision whether the candidate object is indeed a good candidate for being merged with the seed object, the relative border to the seed object (percentage of the border in contact with the seed object) is relevant and not the relative border to the class.

The following is an example of a navigation into the neighborhood of selected seed objects that includes classification of neighbors and eventually a merge with neighbors: (1) hierarchical level n: select starting objects that are potential nuclei with shapes that do not satisfy yet are selected by defining the domain appropriately; (2) hierarchical level $n-1$: navigate to the neighboring objects in a sub-process that fulfills the condition to have, for example, more than 0.5 (more than 50%) relative border to the related starting object that is the PPO 1 (the object where the navigation started, which is defined in the process PPO 1, one hierarchical level higher), and reclassifying them, for example, into *candidate to potentially be merged*. (3) hierarchical level $n-1$: navigate back to the starting object (current image object) and use the process image object fusion; merge the candidate with class *candidate to potentially be merged*, if the classification value for nucleus of the starting object gets improved; (4) hierarchical level $n-1$, neighbor objects: reclassify *candidate to potentially be merged* into the original class, if it has not been merged. This is a very fast process. Through the starting domain, only a few seed objects will be selected for the process, and through re-classification of only a few neighboring candidates, the final decision whether a classification value will improve through a merge will only be made for very few seed-candidate pairs.

In CNL, many tools make the development of a solution very convenient and efficient. This chapter, however, is not meant to be a tutorial, and most of the features cannot be described. Some of the properties are related and similar to other computer languages. Other elements of CNL are very special. Besides the class and process hierarchies and the domain navigation, the following features might be worth mentioning.

3.3.2.4 Maps

The concept of maps allows the processing of several images within one and the same project. Those images could be different ones, for example, when differently stained consecutive slides are presented on different maps, or could be copies of the same image. Even copies of the same image produce several benefits. One benefit is that in an intermediate state of the analysis, some processing experiments or trials might be advantageous. This would, to some extent, possibly destroy the results that have already been achieved. Copying the desired image layers and the object levels onto a separate map allows for auxiliary segmentations without altering the results on the original map. If through this process new relevant objects are found or old ones get improved, they can be copied into the original map. Another benefit is processing speed. Depending on some properties of the objects to be detected, like size and contrast, they can be detected at lower image resolution much faster. If those objects are relatively big, the original map can be copied into a new map with lower image resolution. In particular, for general filter processes on layers where a kernel is used, the gain in performance is dramatic. For half the image size, the total pixel size of the image as well as the kernel size is only 25%. This could lead to a shortening of processing time by a factor 1/16. Therefore, it is always worthwhile to consider object detection on lower image resolution. The objects detected can be copied to the original map and can be adapted relatively fast to the higher resolution.

3.3.2.5   Object variables

Variables can be used for all kinds of CNL elements like for processing parameters, for levels and many other subjects for a more elegant programming style. Besides that, image object variables can be defined and stored into selected image objects. This provides information for several beneficial activities. One purpose of use results from the well-known balance problem of whether to store information or re-calculate it. In case the information is expensive performance-wise for a re-calculation and there is the need for repeated access to these data, it is advantageous to store the data. One example is the calculation of a complex context; another one is the evaluation of the classification value. In case the class description of some objects is complex, the calculation of their classification value is somewhat time consuming, and if this information is used in several processing steps, it helps to store the classification value into the related objects.

Another purpose results from the evolutionary stepwise approach. The state of the object hierarchy changes many times during the analysis. Old states cannot be re-calculated, but they can be stored. By storing states into objects, their evolution can be accessed and it can be evaluated whether their states have been improved or not.

Furthermore, object variables help to make processing local and object-individual. In complex images, usually global processing is not appropriate. For different areas, different processing algorithms might be required. This also can be true for individual objects. In some regions, the same type of object might have less contrast or might be brighter. For their optimization, individual processing might be required. For this, individual information can be collected and stored into an object variable and be used to parameterize object–local processes. A simple example is an object–local threshold segmentation where the thresholds are represented by individual object variables, which might be very different for the different objects.

In the following, an example of context-driven analysis is presented. Usually, there are several ways to Rome and there are also several paths of context-driven analysis to achieve a successful solution. The following example is just one possibility for approaching the task of analyzing tissue slides.

3.4 Example of Context-Driven Analysis

The analysis of *H&E*-stained tissue slides will serve here as an example for demonstrating the principles of context-driven analysis. *H* stand for hematoxylin and *E* for eosin. H mainly stains the cell nucleus, and E mainly cytoplasm and extracellular proteins. H&E stain is the most important stain in histology.

The full package of these particular analysis algorithms, that is, the full CNL program, or how it is also often called: the rule set, for analyzing H&E-stained tissue slides, is not described in detail. The focus is on demonstrating concepts.

Today H&E-stained tissue slides are most commonly used in the clinical practice for diagnosing patients through histopathological investigations. Histopathology, the macroscopic and predominantly the microscopic appraisal of tissue, plays an important role in diagnostics and in particular a dominant role in diagnosing cancer patients. In the microscopic investigation of tissue slides, cells are visible in detail. Their appearance and their arrangements represent decisive information for the pathologist as well as for automated analyses. In many cases, cancer cells look very different compared to normal cells and also their arrangement is different. In normal tissue, epithelial cells might be arranged in forms of glands, representing their function, which is expressed in this geometrical order. In cancerous areas of the tissue, this order might be deteriorated and the cells and their nuclei might be enlarged. In H&E-stained slides, cells, in particular nuclei, are clearly visible in detail and their individual and their networking properties can be studied very precisely. Cell types and their functions are, however, not highlighted through markers like, for example, in immunohistochemistry (IHC) slides. The focus in H&E is on morphology. As mentioned in the gland example, however, cell types and cell functions can indeed be derived from the morphology on the different scales—from subcellular structures to cell arrangements.

For a more advanced and detailed analysis of tissue properties and the investigation of cell–cell interactions, many different methods have been developed. Several ways of highlighting special cell functions or cell types are in use. Among those, IHC is most commonly applied in the clinical routine. In IHC, the ability of certain molecules to bind specifically to special proteins is made use of. For instance, in the functional operation of the immune system, the interaction among and between immune cells and other cells is mainly driven by specific bonds between proteins located on the cell surfaces of the individual cells. For this kind of cell–cell recognition, specific bindings of molecules are required, which means that a specific and stable bond is predominantly achieved between two types of proteins only. In some cases, the constituents of those bonds can also bind to other proteins, but not to too many others. If proteins take part in different bond pairs, usually different processes are started, which means that different pathways are triggered. Through those specific protein–protein interactions, the immune cells can, for example, distinguish own normal cells of the organism from abnormal or foreign cells, self from non-self. This outstanding capability had been developed by nature to a level of sophistication and complexity that goes far beyond what can be created artificially. Nevertheless, the potential for establishing specific bonds is extensively used for diagnostics and for treatment. In IHC, mainly natural antibodies are used for this purpose. If they bind specifically, what they preferably should do, they are called monoclonal antibodies. For making these bonds microscopically visible, other molecules or objects are attached to the antibodies. These attached objects are usually larger than the antibodies and absorb certain frequencies of light and, therefore, carry a color useful for microscopic bright-field investigations. Fluorescent molecules [immunofluorescent (IF) staining] are also common but are today still more widespread in research than in the clinical practice. This might change in the future driven by research and the evolution of digital pathology.

Different types of antibodies that are differently colored can be used to stain one and the same tissue slide. This way in many cases, cell types and their states can be marked unambiguously. For bright-field investigations, up to four different stains can, in some cases, be achieved in a useful way, but in IF staining, four are standard and more than six are possible. For an advanced image analysis for tissue phenomics more than four markers are desirable.

When bright-field microscopy is the method of choice, many different stains cannot all be placed on one slide. The solution here is, for example, to use dual stains (including the counterstain; this represents three different stains) for visualizing co-localizations of proteins on cellular level. For more information, consecutive slides (with multi-stain or not) can be stained with other markers. The consecutive slides can then be co-registered digitally (Schönmeyer et al., 2015; Yigitsoy et al., 2017). If cutting of tissue slides is done carefully, co-registration can lead to close-to-cell-precise co-localization of different proteins. This is possible as the slide thickness is about only three micrometers, which is less than the diameter of a cell. Multi-staining of one slide is required for the cell-precise determination of co-localization, whereas co-registration enables the co-localization analysis within regions, which are not so much larger than the sizes of individual cells. Co-localization within regions is, for example, good enough when the density of T cells needs to be measured within the tumor. In this case, two consecutive tissue slides can be used: one with a tumor marker and the other with a T-cell marker. Through co-registration, one gets an overlay of T cells with the tumor region.

H&E staining is, on one hand, not very specific and, therefore, solving the H&E problem might appear meaningless to many people working in digital pathology. On the other hand, H&E staining is most often used by pathologists, as it delivers valuable information. A wide range of features can be extracted, and immune cells (not the states they are in) are distinguishable from other cells. Furthermore, hematoxylin is nearly always used as a counterstain in IHC when specific proteins such as CD8 (to mark T cells) are stained. The counterstain helps to understand the scenario as a whole. It is useful for a more holistic investigation of tissue slides. It is, for example, valuable for determining ratios of numbers of specific cell or cells in a specific state stained by an antibody to all other cells stained by H. As the H counterstain visualizes nearly all types of nuclei, this is very meaningful. The specifically stained proteins in IHC are less difficult to detect, as their contrast is strong. Therefore, solving the *H&E problem* solves also the most difficult part of IHC slide analysis!

3.4.1 H&E Analysis Problems

The variation of the appearance of structures in H&E-stained tissue slides is huge. Many problems result from overstaining, inadequate tissue cutting, and faulty digitization. In digitization, the biggest problems are out-of-focus regions. As the autofocus of slide scanners is based on relatively large regions and not on individual cells, already a slight warping of the tissue slide can lead to blurred regions in the digital image. Another origin for blurred regions is air bubbles trapped between the two glass slides that sandwich the tissue slide. If those problems are too serious, those regions should be excluded. They, however, need to be detected as part of the image analysis.

Another problem is the shift of the images in the different color channels (red, green, and blue). This can, in principle, be avoided but is nevertheless quite common in the daily routine (see Fig. 3.1).

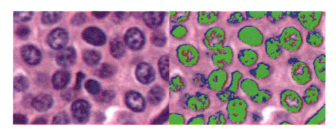

Figure 3.1 Example for how sometimes the different color channels are spatially shifted against each other. In the image itself, this is hard to recognize (left image), but when thresholds are applied, it becomes obvious that objects derived from the red and the blue channels are shifted against each other. Green objects result from thresholds of the blue and green channel, blue objects from the blue channel only, and red from the red channel only. The blue channel is shifted west–north–west.

Cutting artifacts can be very problematic. Experts find a way to deal with this problem. Machines struggle somewhat with strong cutting defects.

Overstaining in H&E affects mainly the green image layer channel. For this reason and also because of the spatial shift of layer channels, the following analysis focusses mainly only on one channel, the red layer. This layer is also reliable for IHC images for analyzing the structures visible in the counterstain.

The expectation is that image analysis deals with all those problems similar to how humans solve those problems. To cover a huge variety of defects, the capabilities of experts are, however, still superior to that of machines. The benefit of rich quantification is, however, so crucial that in the end, the advance of digital analysis will be essential also for the daily routine and, therefore, will probably also lead to a higher quality of sample preparation. People will be willing to pay this minor price for the huge benefit of rich quantification.

3.4.2 Concrete H&E Image Analysis Example

3.4.2.1 Color channel analysis as a starting point

The color tones of H&E images and their contrasts vary considerably. Therefore, to start with a color analysis of an image is meaningful. This analysis can serve as context for the next steps. In the following specific analysis, this is achieved through analyzing the dependency of the value of the quantile of the pixels in the red channel on the quantile-percentage of the pixels. In the quantile algorithm, the pixels are sorted after their brightness. Maneuvering through this sorted list of pixels from dark to bright, their brightness changes faster or slower, depending on how many pixels exist with the same brightness. Taking always the same portion of pixels from the sorted list (starting with the darkest one), the difference of the brightness from portion to portion tells us something about the number of objects in this brightness regime. If the brightness of the next portion is the same as that of the previous one, there are many objects— at least pixels—in this brightness regime. A threshold in this brightness regime would not make much sense. In an analysis in that spirit by additionally making use of the darkest pixel in the image, three thresholds are found. The results for very different images are shown in Fig. 3.2. This is just to get started with the hope to extract some information. It can be seen in Fig. 3.2 that dark regions are separated from less dark ones and from bright regions. The goal is to find dark nuclei, specifically nuclei from immune cells, in the dark regions and other nuclei in the medium bright regions. The bright regions represent background. This is a color channel analysis to start the analysis. Once the first relevant objects are found, also the color analysis can be made more precisely.

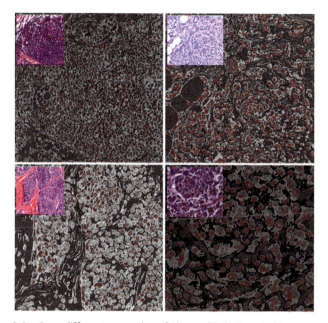

Figure 3.2 Four different examples of the multiple threshold segmentation with threshold derived from the color analysis of each image as described in the text. In the left upper corner of each segmentation example, the de-magnified original image is shown. Dark objects from low thresholds in the red channel appear in brown, medium threshold regions in transparent (bright), and bright ones in grey.

3.4.2.2 Isolated nuclei: first context objects

In context-driven analysis, the path of going through the different processing steps should always be from detecting simple to ever more difficult objects to finally the most complex ones. The lymphocytes are apparently the most suitable objects to start with. The shape of their nucleus is well-defined and different from that of other cells. Also, their contrast is high. They, however, can be confused with nucleoli (as explained later in detail). Nicely shaped ellipsoidal nuclei that are separated from other structures, well-isolated nuclei are also not difficult to detect. If some of them have a relatively good contrast to their environment and have a nice elliptic shape, they might be the easiest objects to detect reliably. There are many different ways to detect them. A simple iteration with varying threshold segmentations can already solve this, when elliptic objects with good contrast are segmented for one particular threshold and are kept during the iteration. In general, when very fast algorithms are chosen, more than one algorithm can be used in a sequential process. The first algorithm might lead to the detection of some nuclei, and the following ones might deliver additional ones. Here, however, we solely use one that is based on the gradient in the red channel at the border of a nucleus.

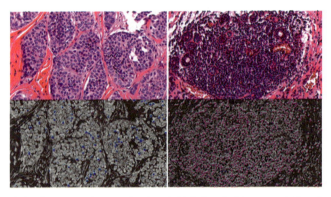

Figure 3.3 Segmentation result after initial threshold segmentation for two different images, right: inflammation and left: tumor region, with original image at the top and segmentation result at the bottom. Although only a few nuclei are found in the tumor region, valuable information can be collected. In the right image, most of the lymphocytes are already detected. This results from an additional growth process of the dark segments into the darkest neighborhood in case they are smaller than expected lymphocytes.

Three different thresholds on the red channel are used: bright to separate the background, medium for normal nuclei, and dark for dark nuclei–like lymphocytes. The brightness and the contrast of the nuclei vary considerably, and we do not expect that great results are achieved by these three thresholds. Nevertheless, there is a good chance that some of the nuclei are segmented right away. If some of the objects resulting from the thresholds have good contrast and the right shape, they get classified as nuclei. Some results are shown in Fig. 3.3. By this procedure, nuclei that are part of a dense cluster cannot be detected, but some of the heterogeneous cancer nuclei that are isolated from other objects can indeed be segmented. Here a first simple context-driven process is applied. For the dark segmentation threshold, only the border of a heterogeneous nucleus and its nucleoli might be segmented. This object does not fit well to a class description of a nucleus. The inner part of such nuclei might be much brighter, but this part can be completely surrounded by the segmented dark nucleus border (Fig. 3.4). Small bright objects with a border solely to medium dark objects serve here as context. They are detected just by their size and brightness. As they potentially might enclose nucleoli, those nucleoli first have to be incorporated into the bright object and next it can be tested whether a merge with the surrounding dark objects leads finally to objects that fit to the class description of a nucleus. This is a very fast process and leads already with a small but reasonable probability to some good results. To grab more of the isolated nuclei, another process follows, where the background is flooded in iteration to stop at ever stronger gradient in this processing example for context-driven nucleus detection.

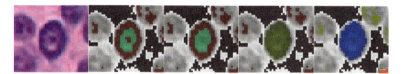

Figure 3.4 Example of a simple context-driven analysis. Relatively small bright or medium bright objects serve as context objects for detecting heterogeneous nuclei. On the very left, the original image is shown, neighbored by the first segmentation and classification result. The small bright object is classified in green as a context object candidate. It encloses a dark object (classified in brown) and is enclosed by another dark object. Going from left to right, the enclosed dark object is merged with the context object and in the next step merged with the enclosing dark object. In the last step (very right image), the resulting object is classified as heterogeneous nucleus (blue).

The color channel analysis leads to a very reasonable result for segmenting the background where no relevant structures are present in this region. The calculation of the threshold is modeled in a way that there is some safety distance of the background brightness to the brightest relevant object. Now, in the applied segmentation process, the background is extended at its borders into regions where the gradients at the border are small. If the gradient locally exceeds a certain value, the growth process stops there. For this procedure, the red channel is used and the threshold for the gradient is lowered in iterations. One could also say the growth process of the background stops at edges—first at weak edges and later in the iteration at strong edges. However, one has to keep in mind that the gradient carries more information as edges do not show in which direction the gradient is positive. At each iteration step, the emerging objects that get surrounded by the growing background are analyzed whether they are nucleus-like in shape and contrast. In Fig. 3.5, results are shown for some examples.

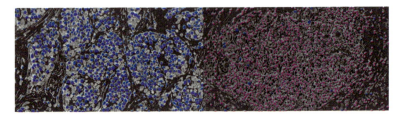

Figure 3.5 Segmentation result after iterative growth of the background. Growth stops at gradients in the red channel with increasing strengths. At each iteration step, the features for the nucleus classification values of the objects enclosed by the background are tested. To save processing time, less complex features are tested first.

From this analysis and the detected nuclei, important information can be extracted that serves as context for the next steps. Contrast and brightness information, for example, about the already detected objects is helpful. Their average brightness might be meaningless as some of them might be very heterogeneous cancer nuclei, which can appear relatively bright on their inside. The brightness of their inner border is, however, more well-defined and significant. This brightness and contrast information is in any case valuable for the detection of the nuclei that are still not found. Also, the hue is helpful for detecting stroma regions as the hue of the stroma is shifted toward red compared to the nuclei. The different types of nuclei can appear quite differently in their brightness. Therefore, it is meaningful to evaluate their contrast, brightness, and patterns in connection with their type before conclusions on the general appearance of nuclei in an image can be drawn.

The heterogeneous nuclei are among the most difficult to segment objects in H&E slides, in particular when they are clustered in close proximity. If some heterogeneous nuclei are found in the first segmentation step, valuable information has been gained. The information about their sheer existence might already trigger more extensive and complex image analysis algorithms and their particular appearance can help to detect similar nuclei. The results of this first segmentation serve as context for the next steps. It influences the further analysis.

There is one more important information that might be gained. The flooding of the background stops in the end at strong edges, which means that also contrast-rich lymphocytes are segmented and detected. Besides them, nucleoli are found. Some of them can be confused with lymphocytes, but the majority of them do not have the right size and shape. Measuring the ratio of the number of nucleoli to the number of lymphocytes can lead to a *nucleolus alarm* giving a hint for the existence of heterogeneous cancer nuclei. This can be used for an extra segmentation process to detect heterogeneous nuclei, which is only performed when the alarm is high enough. In the image shown in Fig 3.6, this is indeed the case.

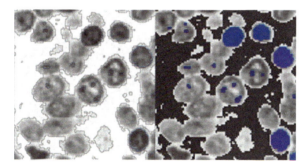

Figure 3.6 Nucleolus alarm. If many nucleoli (dark blue small objects) and/or heterogeneous nuclei are detected in certain regions, an extra processing effort can be triggered for the detection of heterogeneous nuclei.

This does not mean that heterogeneous nuclei are lost in the analysis for images with minor alarm. In the end of the CNL rule set, there will be a template match for those nuclei. This template match is, however, expensive in terms of processing time. It will only be performed in regions where no nuclei have been found yet. The likelihood for large remaining areas used for such kind of template matches is relatively small when the alarm is low. Therefore, the extra processing step triggered by the alarm is just for saving processing time.

3.4.2.3 Some remaining large nuclei by splitting and *inside-out* procedure

If many heterogeneous or large nuclei are detected by the previous processes, an extra process is triggered. This is again to save processing time, as the following procedures are fast but only make sense if the chance for detecting such nuclei is high. Like in the process of Section 3.4.2.1, the heterogeneous nuclei are filled with small enclosed brighter objects. Now those resulting objects represent very likely clusters of heterogeneous nuclei. With standard splitting algorithms, it can be tested whether the split objects could indeed represent nuclei. This is not shown here in detail, as such algorithms are well known. One way of non-standard splitting, however, is used where again small bright objects are used as context like in Section 3.4.2.1. In contrast to the process in Section 3.4.2.1, the process for splitting clusters is here somewhat more sophisticated and is not executed as a merge but rather as a growth process. As shown in Fig. 3.7, the enclosed bright objects grow into the darker ones. First they simply grow without any condition and later with increasing surface tension. In the end, there is a process to finalize their shape by optimizing their contrast.

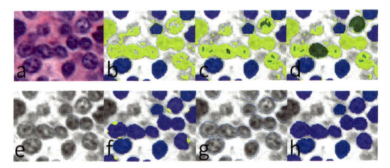

Figure 3.7 Inside-out-separating clusters of heterogeneous nuclei. In (a), a cluster of heterogeneous nuclei is visible. In (b), a threshold segmentation is applied, eventually producing holes within the clusters, which are classified as potential inner parts of heterogeneous nuclei (small green objects) in (c). Larger inner parts are grown first (d) toward the borders of the heterogeneous nuclei. In (e), smaller inner parts are grown and conventional separation technique has been applied for clusters without inner parts. In (f–h), some shape corrections were performed.

3.4.2.4 Nuclei of immune cells

In H&E, one of the most well-defined objects are the nuclei of lymphocytes. They are usually among the smallest and darkest relevant objects, and their shape is simple. They are very small, very dark, and very round objects. There is, in particular, one difficulty. The nucleoli of cancer cells can have, by chance, a very similar appearance (see Fig. 3.8). In cancer cells, the nuclei are altered. They are often enlarged, and they are composed differently than normal nuclei. In H&E, as mentioned before, they often look heterogeneous in a special way. The membranes of the nuclei appear dark, and on their inside, they can show bright and dark clusters. The dark clusters, the nucleoli, can in some cases be confused with lymphocytes. The contexts of nucleoli and lymphocytes, however, are different. Usually for nucleoli, the dark nucleus membrane represents the context, whereas lymphocytes are usually neighbored by several or many other lymphocytes.

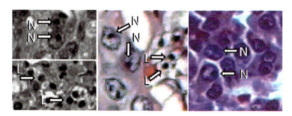

Figure 3.8 Nucleoli–lymphocytes confusion. Nucleoli can have a very similar appearance with respect to shape and color in H&E images as lymphocytes (N, nucleoli; L, lymphocytes).

In this second or third segmentation step, these dark objects of a certain size are detected even when they are clustered. They appear with their roundness rotationally invariant and a simple template match is helpful. In Fig. 3.9, the result of such a template match is shown. A so-called ring filter is applied, where for each pixel in the image, the brightness in layer 1 of all pixels within a certain radius r is averaged and the difference d of this value to a one-pixel-wide ring with a radius $R = r + 1$ is calculated. This value d is plotted for each investigated pixel in the image (Fig. 3.9). If the radius r fits to the radius of the nucleus, d has a maximum value when the investigated pixel is located in the center of a lymphocyte. The story is, however, slightly more complicated as the lymphocytes vary somewhat in size. This mainly results from cutting thin slices of tissue where the nuclei can be hit either in their center or somewhat off-center. Therefore, three different radii are used for each pixel location, three different d are calculated, and the d with the highest value is plotted for each pixel in the image (see Fig. 3.9). As the filter kernel is small due to the small size of the lymphocytes, this filter process is fast. In the center of the lymphocytes, a maximum on the filtered layer appears, which is detected by another filter and serves as a seed for finding the nucleus of the lymphocyte. For each maximum, the best of the three ring templates is known and can be used to grow the seed to the full size of the nucleus. This is not done in a blind fashion, but rather by taking the brightness of the object borders in the red channel into account as well. Results are shown in Fig. 3.9.

Figure 3.9 Template match for lymphocytes: (a) Original image. (b) A threshold segmentation with the value of the threshold derived from lymphocytes that have been detected before. (c) White objects that are classified directly from the threshold segmentation. (d) Additional white objects that are detected through conventional object separation of the dark yellow large objects in (c). (e) Lymphocytes (in red) that had been detected in earlier processes and were stored on a different map and now got copied into the processing map. Classification values of red and white lymphocytes are stored into object variables and on a layer (white) for the decision which detection is best, if there is an overlap of white and red lymphocytes resulting in the segmentation in (f). (g) Strength of the template match and (h) maxima of the matching strength segmented as one pixel sized yellow objects. For the already detected lymphocytes, the maxima represent a double check for the correctness of the segmentation with the expectation that only one maximum is found roughly in the center of the object. In some cases, corrections are made. (i, j) Result of these corrections and the growth process of the pixel sized objects with some contrast optimization in the end. In red are the objects classified as lymphocytes and in blue classified as nuclei, as they are either too big for lymphocytes or not dark enough.

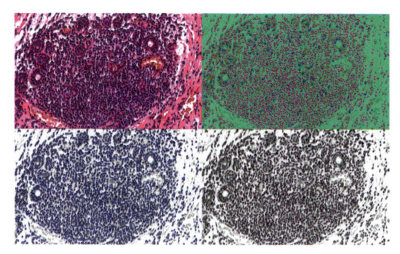

Figure 3.10 Result for inflammatory region. Upper left image is the original image, and the other three images represent different forms of how to present the segmentation result (red, lymphocytes; blue, other nuclei; greenish/yellow, nuclei with low classification value).

3.4.2.5 Template match for heterogeneous nuclei on 10× and 20×

After the described processing steps, quite some knowledge has been gained about a particular image or a particular region within an image. This knowledge can be used for the next and the final processing steps. In case there are still very dark unclassified regions remaining and at the same time several lymphocytes have been detected, extra effort will automatically be put into the search for lymphocytes in those areas. More important than this is a search for the remaining difficult-to-detect nuclei, for example, the heterogeneous nuclei in particular, if they are clustered. For performance reasons, this is executed first at a relatively low resolution of 10× again in the form of applying a ring filter. In case there are cancerous cells within the image, they are probably enlarged and for their detection, this level of resolution might be sufficient. Later the same process will be executed on 20×. As usual most nuclei are already detected at that state of processing, and there are only small regions left for this *expensive* template match. The regions to be considered are furthermore reduced by the pre-selection from the template match on 10×: Only in regions where some match has been detected on 10× will be considered for a template match on 20×. Results are shown in Fig. 3.11. The individual steps are not shown as they are similar to the template match for the lymphocytes shown in Fig. 3.9. In Fig. 3.10, the result is shown for the inflammatory region discussed before. For this example of a rule set, context played an important role. In individual processing steps, certain information could be derived that was used for the next steps. Machine learning algorithms were not used here. In principle after each processing step, training algorithms could be applied that can be used for detecting similar objects in the next step. For performance reasons, this is in particular applicable when whole slides are analyzed and the training is only performed on selected fractions of the slides. Brieu *et al.* (2016) have demonstrated this principle in a broad sense.

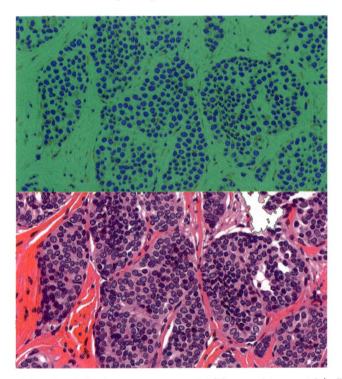

Figure 3.11 Result for heterogeneous nuclei (blue, nuclei; greenish/yellow, nuclei with low classification value).

3.5 Conclusion

Context-driven image analysis delivers several decisive advantages. Most important is the resulting possibility to process very complex images successfully. Through context, the nearly endless number of possibilities for detecting relevant objects gets reduced and manageable. Furthermore, in most cases, context-driven analysis operates on the bases of objects and regions with local processing, which is usually much faster than global processing. CNL enables context-driven image analysis by providing the integration of important techniques and the relevant tools for stepwise advancing processing from simple to complex tasks. In the future, machine learning mechanisms will become more important in the context-driven processing path, and within the next years, ways will be developed how machine learning can help to optimize this path.

In the image analysis example shown here, multiscale structures and hierarchies did not play an essential role. These aspects become more important for a more holistic form of image analysis, where specific super-objects or super-regions of cells come into play. Those objects or regions, like glands or cancer regions, are certainly important for tissue phenomics. As they are much bigger than cells, manual annotations also can solve this problem in some cases. For more complex high-level structures, automation is required. Multiplexing or virtual multiplexing is here very helpful. From the different stains, valuable information, such as the PD-L1–positive tumor regions, can be defined. As tumors can be very complex, the structures of those regions can be complex. It will be discovered through data mining procedures which of those regions will be important in the end. Focusing here on H&E images, this aspect is ignored. As CNL supports hierarchical structures and co-analysis of different images, those aspects as well as longitudinal studies can, however, be covered comprehensively.

Tissue phenomics is a very young discipline, and the importance of context-driven analysis will increase over time. Today new valuable results can already be achieved by measuring cell densities of certain cell types in the tumor and its microenvironment. Precise quantification is difficult to achieve manually and image analysis can improve the situation already with little effort. The full power of tissue phenomics, however, will be unfolded when the social network of cells can be quantified. For this complex task, context-driven image analysis is required.

Acknowledgments

Martin Baatz played a key role here in implementing these first steps and features. The start of a new technique is the most critical phase and requires passion, courage, and phantasy. An year later, Martin was joined by Arno Schäpe, who played a crucial role in the development of CNL in the earlier stages, in implementing CNL as a scripting language in the later stages, and as a software architect for always finding elegant solutions for the implementation of new basic methods. Günter Schmidt introduced several very important CNL elements. He combined pixel processing with object-based processing, paving the way for CNL becoming an integration technique. He also introduced linking of objects (object networks) and the parallel analysis for auxiliary, alternative analyses, which also can be performed on different resolution levels for performance reasons. Maria Athelogou took the first initiatives for the required implementations to process medical images, including three-dimensional scans. In the last 3 years, Nicolas Brieu demonstrated how to perform context-driven analysis by integrating sophisticated machine learning methods. Many important improvements have been made over the years, and many people contributed to the progress that has been made. I am grateful for this and that I had the opportunity to work with those creative minds.

References

Athelogou, M., Schmidt, G., Schäpe, A., Baatz, M., and Binnig, G. (2007). Cognition network technology: A novel multimodal image analysis technique for automatic identification and quantification of biological image contents. In *Imaging Cellular and Molecular Biological Functions* (Springer, Berlin, Heidelberg), pp. 407–422.

Baatz, M., Schäpe, A., Schmidt, G., Athelogou, M., and Binnig, G. (2005). Cognition network technology: Object orientation and fractal topology in biomedical image analysis. Method and applications. In *Fractals in Biology and Medicine* (Birkhäuser Basel), pp. 67–73.

Baatz, M., Zimmermann, J., and Crawford, S. (2008). Detailed quantification of biomarker expression in the context of relevant morphological units. *Cancer Res.*, **68**, 5129–5129.

Baatz, M., Zimmermann, J., and Blackmore, C. G. (2009). Automated analysis and detailed quantification of biomedical images using Definiens Cognition Network Technology. *Comb. Chem. High Throughput Screen.*, **12**, 908–916.

Benz, U. C., Hofmann, P., Willhauck, G., Lingenfelder, I., and Heynen, M. (2004). Multi-resolution, object-oriented fuzzy analysis of remote sensing data for GIS-ready information. *ISPRS J. Photogramm. Remote Sens.*, **58**, 239–258.

Brieu, N., Pauly, O., Zimmermann, J., Binnig, G., and Schmidt, G. (2016). Slide specific models for segmentation of differently stained digital histopathology whole slide images. In *Proc. SPIE 9784, Medical Imaging 2016: Image Processing, 978410* (21 March 2016); doi: 10.1117/12.2208620.

Schäpe, A., Urbani, M., Leiderer, R., and Athelogou, M. (2003). Fraktal hierarchische, prozeß- und objektbasierte Bildanalyse. In *Bildverarbeitung für die Medizin 2003* (Springer, Berlin, Heidelberg), pp. 206–210.

Schönmeyer, R., Brieu, N., Schaadt, N., Feuerhake, F., Schmidt, G., and Binnig, G. (2015). Automated whole slide analysis of differently stained and co-registered tissue sections. In *Bildverarbeitung Für Die Medizin 2015*, H. Handels, T. M. Deserno, H.-P. Meinzer, and T. Tolxdorff (eds.) (Springer, Berlin, Heidelberg), pp. 407–412.

Yigitsoy, M., Schmidt, G., and Binnig, G. (2017). Hierarchical patch-based co-registration of differently stained histopathology slides. In *Medical Imaging 2017: Digital Pathology, Proc. SPIE 2017* (Orlando, FL, USA).

Chapter 4

Machine Learning: A Data-Driven Approach to Image Analysis

Nicolas Brieu,[a] Maximilian Baust,[b] Nathalie Harder,[a] Katharina Nekolla,[a] Armin Meier,[a] and Günter Schmidt[a]

[a]*Definiens AG, Bernhard-Wicki-Strasse 5, 80636 Munich, Germany*
[b]*Chair for Computer Aided Medical Procedures & Augmented Reality (CAMP), Technische Universität München, Munich, Germany*
nbrieu@definiens.com

Machine learning is a branch of computer science attempting to create systems for inference and prediction that can learn from data without requiring the user to define explicit rules. Machine learning has made tremendous progress in the last two decades, especially in the fields of image analysis, natural language processing, and pattern recognition. This development has been fostered by the availability of large datasets for training these systems as well as reasonably priced massive parallel computing architectures, predominantly graphics processing units (GPUs). Besides the ability to derive decision rules automatically, modern machine learning approaches, particularly representation and deep learning methods, are also able to learn discriminative features on their own and thus

---

Tissue Phenomics: Profiling Cancer Patients for Treatment Decisions
Edited by Gerd Binnig, Ralf Huss, and Günter Schmidt
Copyright © 2018 Pan Stanford Publishing Pte. Ltd.
ISBN 978-981-4774-88-8 (Hardcover), 978-1-351-13427-9 (eBook)
www.panstanford.com

represent the *de facto* standard for designing such systems. In this chapter, we provide a brief introduction to machine learning and sketch the development from knowledge-based systems to deep representations. Furthermore, we discuss recent machine learning approaches based on random forests and deep convolutional neural networks as well as their application to the analysis of whole-slide histopathology sections as a basis to tissue phenomics. Finally, we conclude with an outlook on future developments.

4.1 Introduction: From Knowledge-Driven to Data-Driven Systems

In computer vision and more particularly in digital histopathology, machine learning is nowadays a key component of analysis solutions. Typical applications include, but are not limited to, the detection of regions of interest (*e.g.*, tumor metastasis regions, stroma regions) or for the detection, segmentation, and classification of objects of interest (*e.g.*, nuclei, cells, mitosis, glomeruli, and glands). This exhaustive use is motivated by the complexity of the problems above, by the availability of large amounts of raw data, for example, the public TCGA (The Cancer Genome Atlas) database from the NIH, but also by the recent development of efficient algorithms.

Solving classification or detection problems in computer vision requires the ability to distinguish various relevant objects, structures, or regions from undesired ones. Although this task can often be easily achieved by humans, it is not at all obvious how to solve it by a computer. A general approach for solving computer vision problems using machine learning is to design application-specific handcrafted features. Having identified such features, one usually computes them on a set of training images and uses the obtained feature values as input data (training samples) for training a statistical model. The model is trained to discover the mapping between the feature values and the known output values (labels, *e.g.*, 1 if the patch contains nuclei and 0 if not). To put it in a nutshell, the goal of feature extraction is to identify the relevant visual context information and thereby transfer the problem from the pixel space into a feature space where the mapping (decision function) can be more easily learned.

Prior to the recent advances in feature learning and representation learning (*cf*. Fig. 4.1. and Sections 4.3 and 4.4), a lot of research has been focusing on engineering application-specific features, considering prior domain knowledge about the appearance of the classes to be distinguished. For the example of cell and cell nucleus center detection, a popular family of features is the Fast Radial Symmetry (FRS) transform (Loy and Zelinsky, 2002), which enables effective shape modeling of this type of objects. The underlying gradient-driven voting strategy generates maps in which the centers of dark and disk-like objects appear as dark feature points. A classifier can then be employed using the values of these maps as input feature values to differentiate between pixels corresponding to cell centers and pixels corresponding, for example, to stromal tissue. While a very low effort is required to use such well-proven existing features, finding the right set of features can be complex and may require large efforts for novel applications.

To overcome this problem, feature learning and representation learning approaches are used where not only the model but also visual context information can be directly learned from the data. Such approaches, that is, random forests and deep learning, will be detailed in Sections 4.3 and 4.4, respectively. In Section 4.2, we lay the basis for a basic understanding of machine learning.

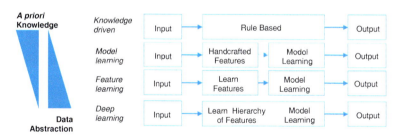

Figure 4.1 Overview of knowledge-driven and machine learning systems, from handcrafted features to representation and deep learning.

4.2 Basics of Machine Learning

Before advanced machine learning methods are discussed in the remainder of this chapter, some foundations are recalled in this section.

4.2.1 Supervised Learning

The first consideration to be made when applying machine learning refers to the intended goal or task. The focus of this chapter is on **supervised machine learning** methods. Their application requires the availability of labeled or annotated data to be used for training. From a mathematical point of view, this means that we assume that there exists a (functional) relationship between the input data samples denoted by x_i and the associated labels denoted by y_i

$$f: x_i \rightarrow y_i$$

where $i = 1, \dots, N$ denotes the number of the respective training sample. The goal of machine learning is to learn a model or approximation $\tilde{f}$ of this often highly nonlinear function f. At this point, the labels y_i can take discrete or continuous values.

4.2.2 Classification and Regression Problems

Depending on the values that the labels y_i attain, one distinguishes two types of supervised learning: *classification* and *regression*. If the labels attain only discrete values, that is, f is taking values in a discrete label space, one is facing a classification problem. A typical classification problem would be to classify a tissue sample as either pathological or healthy. To treat this problem numerically, one assigns discrete values to the individual classes, for example, 1 to pathological cases and 0 to healthy cases. If the labels attain continuous values, that is, f is taking values in a continuous label space, one is facing a regression problem. A regression problem would, for instance, be to regress the probability of a person developing a certain type of cancer based on a set of histological images acquired with several biopsies.

At a first glance, this categorization of supervised learning tasks appears very unambiguous. However, some classification problems can be rephrased as regression problems. This is, first, because some binary classification problems can be solved more effectively if the constraint on the function f attaining discrete values is relaxed. Second, regressing the probability for one label instead of predicting the label itself yields a better representation of uncertainty. An example of such conversion will be provided in Section 4.5 for the detection of cell centers.

To illustrate these observations, we consider the problem of segmenting a digitized histological section into stroma-rich and stroma-poor regions. This task requires the classifier to classify each pixel of the input image with respect to these two categories. The input is a hematoxylin and eosin (H&E)-stained digitized slice and the output a binary image of the same size with stroma-rich regions characterized by label 1 and stroma-poor regions characterized by label 0. The decision of whether a pixel lies in one of such areas is difficult in transition areas, making it in this case useful to regress the probability of one pixel belonging to a stroma-rich region instead of predicting its binary label. It is also important to note that the described classification problem requires the use of visual context information and would be hardly feasible if only individual pixels are regarded as inputs.

Two of the most common tasks in the analysis of digital pathology slides are detection and segmentation. These two tasks are related; for instance, the segmentation of all nuclei in a tissue sample implicitly solves the associated detection problem. There exist some important differences. The detection of an object, such as an individual cell, is usually achieved by highlighting its center with a marker. In contrast, the segmentation of an object requires not only to determine its precise location but also its outline. Both the computational requirements for solving these two problems and the type of analysis enabled by these two problems differ significantly. While solving the detection problem enables the computation of density maps or the study of the spatial interaction between cells, solving the segmentation problem enables more complex analysis based on the morphology of individual cells.

4.2.3 Data Organization

The goal of supervised learning is to find a model that, trained on a set of training samples, is able to perform equally well on a set of unseen samples. This ability is called *generalization*. Learning a model with high generalization performance heavily depends on the quality and organization of the training data as well as on the training setup itself. In this section, we first make some remarks on the generation of appropriate training datasets and second discuss strategies for successful model training while avoiding overfitting.

Training set generation: The generalization performance of a trained model is generally limited by the amount of training examples. To reach a certain level of abstraction, the model needs to be presented enough data containing a reasonable amount of variation during training. In the ideal case, the training data should cover the full range of variability expected in the application domain. In a classification setup, this means covering the full intra-class variability while still highlighting the inter-class differences. This requirement can be addressed from different angles: (1) in the process of training data collection, (2) using *invariant features* in the case of classical machine learning based on handcrafted features, and (3) using *data augmentation* in the case of representation learning and deep learning. First, during training data collection, a broad coverage of the data variability can be enforced, for example, by monitoring basic feature distributions and systematically sampling data from main clusters in such distributions. Also, providing appropriate overviews or summaries of the data for manual review might be beneficial. When using handcrafted features, an increase in the generalization performance is yielded by ensuring that the selected features are invariant with respect to transformations that are irrelevant to the classification task (*e.g.*, translations, rotation, mirroring). In the case of representation learning or deep learning where the features are not explicitly formulated but learned from the data, such invariances can be achieved by augmenting the training data samples. For data augmentation, copies of training data samples are added after applying geometric transformations (*e.g.*, shifting, scaling, mirroring, rotation, elastic warping) or introducing intensity variations, for example. Independently of the chosen machine learning approach, a common good practice is also to normalize the input data that, in the context of the analysis of histopathology images, typically translates into *stain normalization*.

Another relevant property of the training data is the *balancing* of training samples regarding the classes or target values. In the case of unbalanced datasets, that is, certain classes are more frequent than others, a supervised model may tend to optimize the prediction performance for the more frequent classes. To avoid this effect, the training dataset can be balanced by enforcing similar numbers of samples per class, or a uniform distribution of target values in the case of regression. Creating a balanced training set from unbalanced

Basics of Machine Learning | 75

data can be achieved either by subsampling the more frequent classes or by duplicating data from the less frequent classes. A more elegant way of dealing with unbalanced training sets is introducing weights in the trained model itself. For many supervised models, extensions for using class-dependent weighting of training data have been developed.

Training of supervised models requires one to obtain realistic estimates of the current model performance on unseen data. When optimizing model parameters, disjoint datasets are required for adapting the model and evaluating its performance. Otherwise, the model may perfectly adapt to the given data, while its performance on unseen data decreases, which is referred to as overfitting. Note that the degree of overfitting is higher the larger the number of parameters to be optimized with respect to the number of training samples. Common methods for training a supervised model while limiting overfitting are *cross-validation* or *bootstrapping* techniques. Both techniques use data subsets for training and testing multiple models. In n-fold cross-validation, the available data are split into n disjoint subsets, and in each of the n runs, a model is trained on $n-1$ subsets and tested the remaining subset (*cf.* Fig. 4.2a). Thereby, each subset is used for testing exactly once, and the results of the model evaluation are aggregated over all folds. In *Monte Carlo cross-validation*, the data are split randomly at a given ratio into training and test set, which is repeated m times, and the results of the model evaluation are averaged. Combining both methods in a Monte Carlo m times n-fold cross-validation generates m random versions of subdividing the data into n folds (*cf.* Fig. 4.2b). In comparison to cross-validation, bootstrapping methods sample data subsets with replacement. For example, the bootstrap aggregating method creates k different training sets from the overall training set by sampling with replacement, which are used to train k different models. Such models are used as an ensemble method on the independent test set.

To provide a realistic estimate on the performance of the final model trained as described above, an additional validation dataset that has not been touched during training is required. Thus, essentially the overall training data should be split, first, into training and validation data, and, second, the training data are split further into *training* and *test sets*, depending on the favored training setup. When splitting the data into subsets, care needs to be taken that all subsets exhibit the same distribution of essential properties as present in the overall dataset, such as class frequencies or other potentially relevant data subgroups (*e.g.*, patient age, gender, treatment groups). This can be achieved by stratified sampling regarding the relevant features.

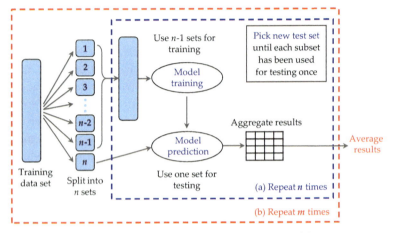

Figure 4.2 *m* times *n*-fold cross-validation with two nested loops repeating training for (a) *n* subsets of the data and (b) *m* repeated splits into *n* subsets.

Table 4.1 Confusion matrix with true positive (TP), false positive (FP), and false negative (FN) predictions per class

	Predicted class			
True class	**1**	**2**	...	**n**
1	**TP(1)**	FN(1), FP(2)	FN(1), FP(...)	FN(1), FP(n)
2	FN(2), FP(1)	**TP(2)**	FN(2), FP(...)	FN(2), FP(n)
...	FN(...), FP(1)	FN(...), FP(2)	**TP(...)**	FN(...), FP(n)
n	FN(n), FP(1)	FN(n), FP(2)	FN(n), FP(...)	**TP(n)**

Evaluation of the model performance during and after training requires appropriate metrics. In supervised learning, the most straightforward metric is the overall accuracy (true predictions/all predictions) of the prediction with respect to the true outcome. However, depending on the studied problem, more specific metrics might be better suited, such as average accuracies, sensitivity, specificity, positive and negative predictive values, f1 score, and ROC curves. For multi-class classification, the confusion matrix provides a comprehensive summary of the classifier performance (see Table 4.1).

4.3 Random Forests for Feature Learning

One of the main disadvantages of classical machine learning based on handcrafted features is that new sets of application-specific features must be designed for each newly encountered problem. A straightforward approach to jointly learn both the features and the model consists of automatically selecting the best set of representative features on infinite and generic feature spaces. As we will detail in this paragraph, this has been successfully performed using *random forests* models with a high number of low-level and application generic visual context features.

In the first paragraph, we detail the divide-and-conquer strategy of the decision tree model with application to classification and regression problems. In the second and third paragraphs, we briefly discuss the interest of ensemble approaches as well as the effect of the different parameters on the modeling and generalization capabilities of a random forests model. The fourth paragraph introduces two families of generic visual features. Finally, we present in the last paragraph two classification and regression forest models for the segmentation of nuclear regions and the detection of nuclei centers in digital histopathology images.

4.3.1 Decision Trees

4.3.1.1 Definition

Let us consider an input sample, that is, usually a pixel in the context of visual scene understanding. A decision tree encodes the routes guiding such input sample from the *entry node* (the root) at the top of the tree to one of the *terminal nodes* (leafs) at the bottom of the tree depending on its successive responses to a series of questions (decision functions D_i) asked at each visited *internal node* N_i. The routes are binary, that is, when a sample s reaches an internal node

N_i, it is routed to the right part of the tree if the response of the sample to the associated decision function is positive $D_i(s) \geq 0$ and to the left part of the tree, if it is negative $D_i(s) < 0$. This is repeated until the sample reaches one of the leaf nodes. The decision tree encodes a hierarchical piecewise model: The internal nodes encode the routing processes that yield the hierarchical partitioning of the samples into subsets where each subset corresponds to a different model that is contained in the leaf node. In classification, leaf nodes typically contain probability distributions of the classes to be predicted. In regression, leaf nodes typically contain posteriors over the continuous variable(s) to be predicted.

4.3.1.2 Decision function

Training a decision tree consists of learning the different routes that best split the training samples into homogeneous subsets with respect to their associated labels. The hierarchical routes are encoded by the different decision functions D_i associated with the internal nodes N_i. An axis-aligned decision function is commonly used because of its simplicity. It is defined as $D_i(s) = F_i(s) - T_i$, where $F_i(s)$ is the response of a sample s to a selected visual feature F_i and T_i is a scalar threshold.

The prevailing training approach follows a greedy strategy, in which a locally optimal choice is made at each node with the hope of finding a global optimum for the tree. Given a subset S_i of training samples arriving at a node N_i, the decision function D_i is estimated by selecting among a set of T pair candidates $(F_t, T_t)_{1 \leq t \leq T}$ the pair (F_i, T_i) that maximizes an entropy-based training objective function $I(t, S_i)$ over the samples S_i.

4.3.1.3 Extremely randomized trees

For low-dimensional feature spaces, pair candidates are typically generated for each of the available features. In the first scenario, a unique arbitrary threshold value is considered per tested feature: This, for instance, includes the mean of the means of the feature values of the samples over the classes in classification or the average of the continuous labels of the samples in regression. Alternatively, each feature is associated with multiple threshold values that are randomly sampled or equally spaced among the feature values displayed by S_i. In this case, multiple candidate functions are tested per feature. Since testing each of the features becomes intractable for high-dimensional feature spaces, it is common practice to randomly select at each node a discrete subset of features to be tested. This is called *randomized node optimization*. If both attributes of the decision function, that is, the tested features and the tested threshold values, are randomly picked, one then talks about *extremely randomized trees*.

4.3.1.4 Training objective function

The selection of the decision function at each node requires the definition of what a good split is. The *information gain* resulting from splitting the incoming sample set S_i into some right and left subsets $S_i^{t,r}$ and $S_i^{t,l}$ is commonly used as the objective function to be maximized. It reads as follows:

$$I(t,S_i) = H(S_i) - \left(\frac{\left|S_i^{t,r}\right|}{\left|S_i\right|} H\left(S_i^{t,r}\right) + \frac{\left|S_i^{t,l}\right|}{\left|S_i\right|} H\left(S_i^{t,l}\right) \right)$$

In classification, $H(S)$ denotes the Shannon entropy. The less random the current class distribution becomes after splitting, that is, the more it follows the known class labels, the smaller the entropies $H(s_i^{t,l})$ and $H(s_i^{t,r})$ become and the higher $I(t,S_i)$ becomes. In regression, $H(S)$ denotes the least square regression error, that is, the trace of the covariance matrix. The less the continuous values associated with the samples $S_i^{t,r}$ and $S_i^{t,l}$ deviate from their respective means, the smaller the regression errors and the higher $I(t,S_i)$. In practice, the information gain is often estimated on a random subset of the samples that arrive at each node. The number of feature values to be computed is, therefore, limited, and the training becomes tractable even on a large training dataset.

4.3.2 Random Forests Ensemble (Bagging)

A random forests model consists of *multiple decision trees* that are trained in parallel on random subsets (bags) of the training set. The underlying idea is that overfitting can be reduced by selecting parameters that are not specialized on the complete training set but selecting independently on multiple training subsets. At prediction, the outputs of the multiple decision trees are aggregated to form a combined output. One of the most common aggregation approaches, which is also used in the two example applications below, averages the class probability distributions for classification and the posteriors over the continuous variable(s) for regression.

4.3.3 Model Parameters

We briefly discuss here the influence of the parameters associated with a random forests model and to the prevailing greedy training strategy.

The forest size/number of trees: The bigger the forest, the more randomness is included through bagging, the less correlated the trees are, and the better the model can generalize to variation within input data. An input sample is forwarded through each of the trees at prediction, yielding a proportional increase in the prediction time with the number of trees.

The maximum depth: The deeper the tree, the more parameters it contains. Given enough training data, a high depth makes it possible to model more complex problems. However, this can lead to overfitting if the trees are too deep compared to the available amount of training data. In this case, it is recommended to include a data augmentation step during training. Adding one leaf level at depth n leads to the inclusion of 2^n additional nodes. This can result in an exponential increase in the training time: 2^n additional decision functions are estimated. Prediction time is only linearly affected as, for a given input sample, only a single additional decision needs to be taken.

The minimal number of samples for a node not to be considered a leaf: This parameter impacts the effective depth of the trees. The higher its value, the less likely it is for a tree to reach its maximal depth.

The number of tries for finding the best feature and threshold attributes of each decision function: The higher the number of tries, the more a forest can explore the parameter space during the selection of the best decision functions, and the better the forest can learn the features that are effectively associated with the problem to be modeled. This parameter linearly affects the training time, while the prediction time is not affected.

The maximum number of samples used to estimate the decision functions at each node: The more samples are considered, the closer the approximate information gain value gets to its actual value. This parameter linearly affects the training time for nodes with many incoming samples, that is, usually the nodes with small depth. Deep nodes usually have less samples and are, therefore, less affected. Prediction time is not affected by this parameter.

4.3.4   Application Generic Visual Context Features

The use of generic visual context features makes random forests out-of-the-box systems in which these features do not have to be adapted for each new specific application. In this chapter, we will focus solely on the *Haar-like features* and *Gabor features*.

4.3.4.1   Haar-like features

Let us consider a pixel p_s corresponding to a sample s. Haar-like features compute relative visual cues at different offset locations and scales around the pixel p_s. Following the notations used in Peter *et al.* (2015), a *Haar feature* λ is defined by four scale-related parameters (v_1, v_2, s_1, s_2) and three categorical parameters (c_1, c_2, w):

$$\lambda = (v_1, v_2, s_1, s_2, c_1, c_2, w)$$

where (s_1, s_2) set the dimensions of two boxes B_1 and B_2 located at $p + v_1$ and $p + v_2$, respectively, (c_1, c_2) set the channel indexes over which one of the expressions given below is computed depending on the value of w:

$$I_1, I_1 - I_2, H(I_1 - I_2), |I_1 - I_2|, I_1 + I_2$$

I_1 denotes the mean intensity of channel c_1 in the box B_1, I_2 the mean intensity of channel c_2 in the box B_2, and H(.) is the Heaviside function. These expressions were further extended into the family of long-range features (Criminisi *et al.*, 2009), to include more complex descriptors such as, for instance, histogram of oriented gradients (HOG).

4.3.4.2 Gabor features

Gabor features encode the spatial appearance as the response to frequency and orientation filters. A Gabor filter is parameterized by three *envelope parameters* (γ,θ,σ), which set the spatial aspect ratio, the orientation, and the standard deviation of the Gaussian envelope, and by two *frequency parameters* (λ,ψ), which set the wavelength of the sinusoidal factor and the phase offset of the sinusoidal wave. A Gabor kernel is defined as follows:

$$G(\gamma,\theta,\sigma,\lambda,\psi) = \exp\left(-\frac{x'^2 + \gamma^2 y'^2}{2\sigma^2} \right) \exp\left(i\left(2\pi\frac{x'}{\lambda} + \psi \right) \right)$$

Given an image channel of index c, the Gabor feature associated to the sample s reads as the result of the convolution $I_c * G(\gamma,\theta,\sigma,\lambda,\psi)$ (p_s) at the pixel p_s. It is important to note that while the use of integral images enables the fast computation of Haar-like features, this is not the case for Gabor features. However, the number of Gabor layers that are generated at prediction time can be limited by training a first decision tree with a first extensive Gabor parameter space and keeping only the subset of parameters that have been selected most often during the node optimization. Only the selected parameter sets are then used to train the second decision tree. To enable selection, each node N_i is associated with an importance weight $w_i = 2^{d_i}$, where d_i denotes the depth of the node. The weights are cumulated for each parameter vector over the nodes at which it was picked as part of the decision function.

4.3.5 Application to the Analysis of Digital Pathology Images

Due to their numerous favorable properties, random forests have been successfully applied in many computer vision related problems. As an example, in Peter *et al.* (2015), classification forests are employed on MR, three-dimensional ultrasound as well as histological data without the features being explicitly designed for each of these applications, thereby demonstrating the autonomous adaptation of this approach to different visual problems.

4.3.5.1 On-the-fly learning of slide-specific random forests models

One of the benefits of random forests models over other prevailing machine learning models such as support vector machines is their high speed at training and at prediction. This motivated in (Brieu *et al.*, 2016) the learning of slide-specific visual context classification random forests models for the detection of nuclear regions through an auto-adaptive strategy. One of the most critical challenges to the analysis of whole-slide digital pathology images is their high discrepancy in visual appearance originating from the use of different cutting protocols, the use of different apparatus and stain combinations for staining, and the use of different scanners for digitalization of the slides into images. The resulting variations are illustrated in Fig. 4.3 in H&E-stained images. We account for such variations using posterior layers generated by random forests models trained independently on each slide based on objects detected using standard blob detection algorithms and selected based on *a priori* known morphological characteristics. The approach is only briefly described here. We refer the reader to the original paper (Brieu *et al.*, 2016) for a more precise methodological description as well as for more extensive qualitative and quantitative results. Given an input slide, the analysis workflow reads as follows.

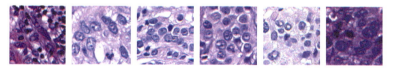

Figure 4.3 Variability of visual appearance of tissue and cell nuclei in H&E-stained images.

First, a set of representative tiled square regions are selected based on the amount of stable background regions, using the maximally stable extremal regions (MSER) detector, and on color information, using clustering of the RGB space. Second, objects of interest are detected on these selected regions: Background is detected as stable bright regions and nuclei as contrasted dark blob objects with surface area and sphericity within *a priori* known bounds. Sphericity is defined as the product between shape index and elliptic fit and, therefore, favors round-shaped objects. MSER and difference of Gaussian are employed as blob detectors, yielding the detection of well-defined nuclei only. Choosing strict constraints on the geometry of these objects yields the selection of very specific and well-defined objects only. The next steps aim at enabling the detection of objects that are not so well defined, for example, the cluttered cell nuclei that do not fulfill the *a priori* known geometric constraints after segmentation. A random forests model with Haar-like features is trained on the well-differentiated objects, that is, cell nuclei and stromal regions, and the corresponding class posteriors are estimated on the full slides in a tile-by-tile fashion.

As shown in Fig. 4.4, both well-defined and cluttered cell nuclei are correctly detected with high posterior values for images with very different visual appearance. This illustrates how the visual context appearance classification model that has been learned on the well-differentiated objects can be successfully transferred to the more difficult cluttered objects and how it can autonomously adapt to the visual appearance of each slide. The same workflow can be extended to create posterior maps for multiple "cell nuclei" classes. The well-defined cell nuclei are, for instance, classified into textured and homogeneous before being used as training samples.

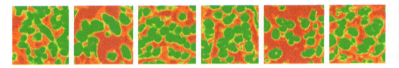

Figure 4.4 Class posterior maps for the class "cell nuclei," computed on the H&E images displayed in Fig. 4.3.

As shown in Fig. 4.5, the prediction step yields the joint generation of textured cell nuclei posterior map and of homogeneous cell nuclei posterior map, therefore displaying the interaction between the two types of cells.

4.3.5.2 Area-guided distance-based detection of cell centers

We have seen in the previous section how classification random forests can be used to learn the visual context associated with nuclear regions. In this section, we briefly discuss on how *regression random forests* can be employed for the detection of cell centers and nuclei centers through the prediction of nuclear distance maps

(Kainz *et al.*, 2015) and present an extension based on the prediction of nuclear surface area maps (Brieu and Schmidt, 2017).

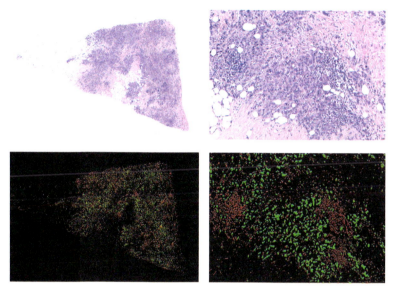

Figure 4.5 Whole-slide image of an H&E-stained slide (top). Class posterior maps indicating the membership probability to homogeneous (red) and textured (green) cell nuclei (bottom).

A regression forest is trained to learn the mapping between local visual context represented by Haar-features and a so-called *nuclear proximity map*, which is a function of the distance map to annotated cell centers (Kainz *et al.*, 2015) (*cf.* Fig. 4.6). Given an unseen image and the trained model, the proximity map with respect to nuclei is predicted. Cell centers are detected as *local maxima* given a kernel corresponding to the *a priori* known radius of the nuclei to be detected. In Brieu and Schmidt (2017), we propose an extension of this workflow by introducing a second regression random forests that is trained on the surface area of annotated cells. As illustrated in Fig. 4.7, given an unseen image, the *nuclear surface area map* is predicted and drives the pixel-by-pixel adaptation of the kernel size used in the local maxima detection step. As before, we refer the reader to the original paper (Brieu and Schmidt, 2017) for a more precise methodological description as well as for more extensive qualitative and quantitative results.

86 | Machine Learning

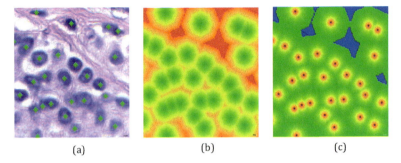

Figure 4.6 Create training data for cell center detection: (a) manually annotated cell centers; (b) distance map; (c) proximity map.

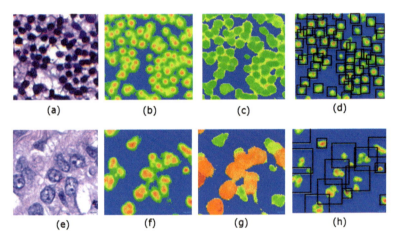

Figure 4.7 Applying trained models: (a and e) input RGB images; (b and f) predicted proximity maps; (c and g) predicted surface area maps; (d and h) applying local maxima detection.

4.4 Deep Convolutional Neural Networks

In contrast to many other machine learning techniques, deep convolutional neural networks have the advantage of being able to learn both the necessary decision rules and a hierarchy of features to be used. Due to the tremendous number of achievements in recent years, convolutional neural networks are an actively researched topic, and it is far beyond the scope of this chapter to discuss all developments in detail. This chapter starts with a brief history of deep learning in Section 4.4.1 and details the main building blocks of modern networks in Section 4.4.2. Finally, some application examples are provided in Section 4.4.3.

4.4.1 History: From Perceptrons and the XOR Problem to Deep Networks

Although the roots of artificial intelligence can be traced back much further, it is reasonable to start this section with the invention of the perceptron algorithm in 1957 by Frank Rosenblatt (Rosenblatt, 1958). In its original form, the *perceptron* is a linear classifier of the form

$$g(x_i) = \gamma(w \cdot x_i + b) \qquad (4.1)$$

where

$$\gamma(x) = \begin{cases} 1, & x > 0 \\ 0, & x \leq 0 \end{cases}$$

is a nonlinear *activation function*. Being a biology-inspired architecture, it mimics the behavior of a neuron aggregating information via its dendrites and firing if "activated" (*cf.* Fig. 4.8).

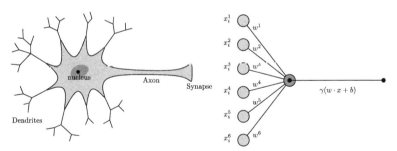

Figure 4.8 Perceptron mimicking the function of a neuron, that is, the simplest version of a neural network.

At the time of its invention, the perceptron was highly promoted by its inventor but created a lot of controversy in the years to follow. The main reason was the observation that perceptrons in their original form are not able to solve the *XOR problem* as depicted in Fig. 4.9. This observation has become popular due to the book

Perceptrons by Marvin Minsky and Seymour Papert in 1969 (Minsky and Papert, 1972). Although it turned out that nonlinear classification problems, such as the XOR problem, could be solved by concatenating two perceptrons, this observation caused a major loss of interest in AI as well as significant budget cutbacks for more than 10 years. This period is commonly referred to as the (first) AI winter. Although it was proven already in 1989 that such a concatenation of two perceptrons endowed with continuous sigmoidal activation functions (*i.e.*, a two-layer neural network) is able to approximate a large class of nonlinear functions arbitrary well, it took almost another two decades for neural networks reaching today's popularity. One significant step toward this popularity was the introduction of convolutional neural networks by Lecun *et al.* (1998), which enabled their use for the recognition of two-dimensional patterns.

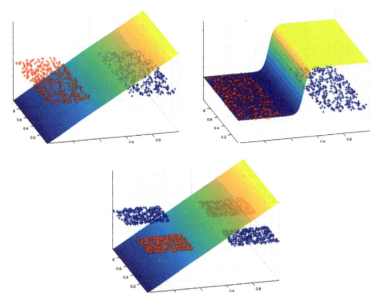

Figure 4.9 Linearly separable point sets (top left), linearly separable point sets with activation function applied to separating hyperplane (top right) and nonlinearly separable point sets associated to the XOR problem (bottom).

The reason for this development was not only that other machine learning techniques, such as support vector machines, boosting, and random forests, became more popular in the meantime, but also that for neural networks to outperform these techniques, several technological advancements were necessary. The first major component is, for sure, the publication of the backpropagation algorithm for training multilayer neural networks by Rumelhart *et al.* (1986). This algorithm facilitates an efficient training of neural networks by back-propagating errors via exploiting the chain rule for differentiating multivariate nested functions.

The second major component is the availability of large online databases containing millions of annotated images that can be used for training. Having such amounts of data at hand, it has become feasible to train very deep networks with millions of parameters. The third reason refers to the rapid development of low-cost but powerful general-purpose graphics processing units (GPGPUs) during the last two decades, facilitating the training of large networks in a reasonable amount of time, that is, less than 1 week. These three major advancements were, of course, accompanied by several theoretical improvements and discoveries for making the training process faster (ReLU-layers) or more robust (dropout layers). However, an aspect that cannot be overestimated is the fact that all this technology is available to a large group of researchers, practitioners, and companies due to online databases being publicly accessible, programming frameworks being open source, computing hardware being affordable, and research articles being published in open access.

4.4.2 Building Blocks

Thanks to the universal approximation theorem, it is known that a concatenation of two perceptrons can approximate almost any nonlinear function f. The problem is, however, that the proof of this theorem is not constructive, that is, it does not provide a strategy for designing a good network. Although a lot of best practices exist and many theoretically motivated improvements have been made, a holistic theory is still missing. In this section, we thus discuss the main ingredients for designing modern convolutional neural networks.

4.4.2.1 Convolutional layers

Convolutional layers are the most important components of convolutional neural networks. To understand what a convolutional layer is, one has to recall the function of a normal perceptron, a two-layer perceptron, and a *convolutional two-layer* perceptron first. Figure 4.10 depicts a standard perceptron on the left. All input variables are connected to one output. Each connection (indicated by a solid line) carries a weight, which is used for computing the weighted average at the output. A *two-layer perceptron* (Fig. 4.10 middle) essentially iterates this concept: The input variables are processed by multiple perceptrons whose outputs form the input of another perceptron. The nodes in the middle form the *hidden layer*. It is important to notice that all nodes of the hidden layer are connected to all input variables; one thus calls the second layer of nodes a fully connected layer. If the input consists of many variables, it requires quite some computational effort to obtain the results for the second layer. A compromise would be to connect each node of the second layer to only a few nodes of the input layer as shown in Fig. 4.10 on the right.

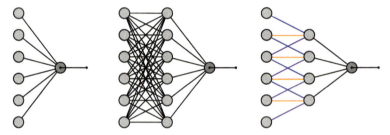

Figure 4.10 Standard perceptron (left), two-layer perceptron (middle), and two-layer convolutional perceptron (right).

4.4.2.2 Convolutional neural networks

Convolutional neural networks go even one step further: They connect the nodes of the hidden layer not only to a fixed number of input variables but also share their weights as indicated by the colored connections. This means that the second layer is created via sliding one perceptron over the input. Mathematically speaking, they compute the second layer by means of a convolution of the input signal with a filter mask carrying the shared weights. The set of filter weights is often called feature, and the result of such a convolution is, therefore, called a *feature channel*. In practice, convolutional layers feature another important difference compared to the examples shown in Fig. 4.10: They use multiple filter masks in order to create not one but multiple outputs. In the case of processing image data, this means that the output of one convolutional layer may be three dimensional as indicated in Fig. 4.11. The subsequent convolutional layer can then be composed of three-dimensional filter masks integrating not only information in a small spatial neighborhood but also from various feature channels.

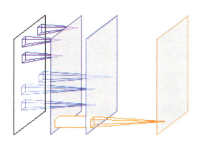

Figure 4.11 Three-dimensional convolutions (orange) accumulate the weighted average of multiple two-dimensional filter maps (blue and purple).

4.4.2.3 Loss functions

Loss functions are not a part of the neural network itself, but they are inevitable for network training as they facilitate to measure how well f is approximated. A popular example is the *L2 loss*,

$$\frac{1}{2}\sum_{i=1}^{N}(f(x_i)-y_i)^2$$

which penalizes the sum of squared differences between the predicted label and the true label. As the backpropagation algorithm is a derivative-based optimization technique, it is beneficial if the loss is differentiable. Throughout the last decades, a plentitude of various loss functions has been proposed to achieve various goals such as robustness to outliers, for example. As the choice of the loss function depends very much on the application and the associated goals, this chapter does not include a detailed review of all possibilities.

4.4.2.4 Activation functions

Activation functions are another important ingredient of any convolutional neural network. The output of each perceptron is fed to a nonlinear activation function. Incorporating nonlinear activations is crucial for being able to approximate nonlinear functions f. The equation describing the perceptron employs a non-differentiable step function, differentiable counterparts such as the *logistic sigmoid* (*cf.* Fig. 4.12) are used in practice. The reason is that the backpropagation algorithm requires the derivative of the optimized loss function with respect to each weight. The logistic sigmoid has been used extensively for several decades until Vinod and Hinton (2010) propagated the usage of the *rectified linear unit*, denoted by ReLU, and their differentiable approximation. This activation function has the advantage of having a larger derivative, which is particularly useful for training very deep neural networks via the backpropagation algorithm. As such networks consist of many activation functions applied in a nested fashion, the derivatives of weights at the beginning of a network get scaled by large powers of the derivatives of the employed activation function. Thus, activation functions having small derivatives hamper the training of deep neural networks significantly. Today, a lot of research is spent on finding even better activation functions, but the ReLU can be seen as the first of its kinds facilitating very deep neural networks.

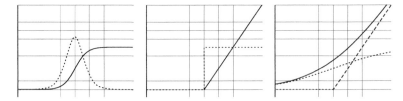

Figure 4.12   Different activation functions.

4.4.2.5   Pooling layers

Pooling layers are another important component of convolutional neural networks. Particularly, in the case of object recognition, it is crucial to design systems that feature certain invariances. Considering the application of mitosis detection, for example, the algorithm should be invariant with respect to illumination changes or out-of-focus blur introduced by the scanner. The algorithm should also be invariant to color variations introduced by inconsistencies during the staining process. Furthermore, the algorithm needs to be invariant to certain geometrical transformations, such as translations, rotations, and scale changes. Some of these invariances can be learned by the system itself, given enough training data are available, or artificially ensured by means of training data augmentation. Concerning geometrical transformations, achieving translation invariance is one of the most important invariances. After all, the recognition system should not care about at which position in an image the mitosis event appears. Ensuring this kind of invariance is the purpose of pooling layers: After a convolutional layer has generated a set of feature maps, the feature responses in a small neighborhood get pooled and eventually replaced by a single value resulting in a feature map of at least half its original size. Employing several pooling layers helps to condense the input information into a single assertion, for example, whether or not a mitosis event is visible in an image. One of the most common types of pooling is max pooling over a 2 × 2 window, that is, one takes the maximum of four neighboring values resulting in a feature map of half the original size.

4.4.2.6 Dropout layers

Improving generalization, that is, achieving a low error rate on the test data, is a very important aspect of machine learning in general and particularly in the case of convolutional neural networks. Models that generalize well do not suffer from overfitting as discussed. As convolutional neural networks may have millions of parameters, the risk of overfitting is thus relatively high. To prevent a network form overfitting, several strategies have been suggested among which dropout layers are among the most popular ones. Dropout layers randomly deactivate a certain number of neurons and their connections during training. This way, one samples from several *thinned networks*. During testing, one simply uses the full network, just with small weights, which results in averaging the results of several thinned networks. This procedure is conceptually similar to ensemble learning techniques such as the random forests discussed above, in that it forces neurons to learn different aspects of the data and thus helps to ensure better generalization.

4.4.3 Application Examples

Mitosis detection in breast cancer histology images with deep neural networks by Cireşan *et al.* (2013) was one of the first works of deep learning for digital pathology applications. It is conceptually similar to the AlexNet introduced by Krizhevsky *et al.* (2012) but does not use dropout layers and ReLU activation functions. However, it can be regarded as a pioneering work for the usage of max-pooling layers. Another important aspect of this work is that it employs a rather small network compared to recently proposed networks, which is possible as the network is applied only to patches extracted from the original image data. This is a typical scenario in tissue phenomics as many structures of interest are comparatively small in contrast to the acquired images; a digitized whole-slide image (WSI) can easily have a size of $100,000 \times 100,000$ pixels, while the extent of a nucleus is typically of 25×25 pixels at 0.22 µm per pixel resolution. Feeding an entire WSI to a neural network would be sub-optimal for various reasons. First, networks taking input data of this size would require a lot of computational resources for training. Second, it would be very hard to acquire the required amount of training examples as each WSI needs to be prepared by a pathologist. In contrast to this, Cireşan *et al.* used only five different biopsy slides for training their classifier, which takes patches of 101×101 pixels as input. Thus, such a patch-wise approach is not only very efficient but also comprises great potential for parallelization as each patch can be processed independently. Such a procedure is, of course, only applicable for detecting structures that are relatively small in comparison to the acquired image.

Deep learning–based cancer metastases detection by Wang *et al.* (2016) has been performed during the IEEE International Symposium on Biomedical Imaging (ISBI2016) Camelyon challenge on cancer metastasis detection in lymph nodes. Participation in this workshop included two tasks, that is, tumor localization and WSI classification. The employed training data comprises 400 WSIs with $40 \times$ magnification. For metastasis localization, the authors employed the GoogLeNet architecture prosed by Szegedy *et al.* (2015), which has 27 layers and approximately six million parameters. From each WSI, 1000 positive and 1000 negative patches with a dimension 256×256 pixels have been selected, resulting in roughly 350,000 patches for training a network for metastasis localization, which achieved an area-under-the-curve value of 0.9234. However, this work is not only noteworthy due to its impressive segmentation performance. It illustrates not only how well convolutional neural networks work in case enough training data are available, but also shows that other classifiers have to be used for achieving the second task of the Camelyon challenge, that is, the classification of WSIs into healthy and pathological ones. For this task, the authors generated metastasis heatmaps using the trained network and computed a set of 28 handcrafted features, such as the longest axis of the tumor region or the percentage of tumor region over the whole tissue region, for training a random forests–based classifier. Consequently, this work can be considered as both: a hybrid approach to WSI classification and a good example how modern classification techniques can be successfully combined to achieve state-of-the-art classification performance.

Identifying tumor regions in cytokeratin (CK) immunofluorescence: In the context of immunotherapies, it is crucial to enable the analysis of the spatial interaction of tumor regions and immune cells and thus to achieve an accurate segmentation of tumor regions. In this example, we show how the use of a convolutional neural network enables the separation of tumor regions from other regions showing similar intensity values in the CK channel but different visual patterns. While immunofluorescence usually offers a higher signal-to-noise ratio than bright-field imaging, the presence of artifacts (*e.g.*, tissue folds, staining artifact) and the low specificity of the CK binding regarding other biological components (*e.g.*, necrotic tissue) make the accurate detection of tumor regions challenging. We consider a two-class problem: tumor regions versus other (*e.g.*, negative, artifact, necrosis), which corresponds to a classical foreground versus background problem. Examples of visual patterns encountered in both classes are displayed in Fig. 4.13. A convolutional neural network is trained on examples of the two classes. We then apply the trained network using a sliding window approach. Results on unseen data are displayed in Fig. 4.14: Posterior heatmaps are computed for each of the two classes and can be further used for the segmentation of the tumor regions.

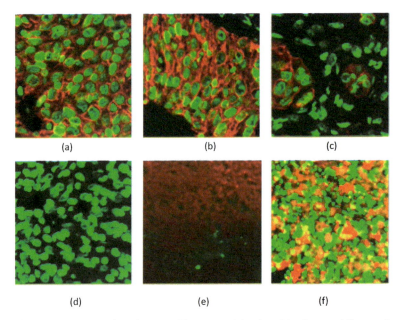

Figure 4.13 Examples of some of foreground (top) and background (bottom) regions used for training the detection of tumor regions. Red and green channels correspond to the cytokeratin (CK) and Hoechst fluorescent channels, respectively. Both foreground and background regions are characterized by a high variability of possible visual patterns: (d) negative CK region, (e) staining artifact, (f) necrosis.

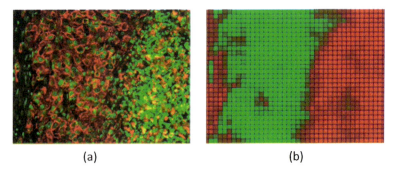

Figure 4.14 Result of predicting a convolutional neural network trained for the detection of tumor region on an immunofluorescence image displaying necrotic tissue regions. (a) Original image with CK and Hoechst fluorescent channels; (b) likelihood of foreground (green) and background regions (red). Prediction stride is defined by the overlapping grid.

4.5 Discussion and Conclusion

4.5.1 Model Complexity and Data Availability

Machine learning and deep learning are well-suited approaches in cases where it is challenging to define explicit features and rules to solve a problem. However, it is good practice while designing a machine learning system not only to consider the complexity of the problem but also the amount of data available to train, test, and validate it. On the one hand, deep models enable the recognition of complex patterns without explicitly formulating prior knowledge. The aforementioned ISBI Camelyon challenge proved that very deep network architectures containing millions of parameters can achieve close-to-perfect accuracies on the difficult problem of automated detection of metastases in H&E-stained WSIs of lymph node sections. On the other hand, deep networks can only reach high generalization performance if they are trained on sufficiently large datasets (also requiring enormous computational capabilities). For example, the data of the Camelyon challenge contain a total of 400 exhaustively annotated WSIs among which one-third is kept for validation purposes. Models containing too many parameters compared to the amount of training data tend to overfit, that is, they typically exhibit significantly worse performance on unseen samples. If few samples are available, less complex machine learning models such as shallow network architectures, random forests, or rule-based systems are likely to generalize better. Thus, such more classical approaches are still preferable in the case of limited training data and if computational efficiency is required compared to very deep network architectures.

4.5.2 Knowledge and Data Teaming

An interesting idea to compensate for the weakness of machine learning approaches with respect to a lack of training data is to team knowledge-driven and data-driven representations. As it is not necessary to learn what is already known, focusing the feature and representation learning only on what is unknown makes it possible to build relatively simple models that can be trained on a few samples while still being able to generalize well. Prior knowledge can be encoded in application-specific features as well as in the architecture of the models or in the structure of the training data. As an example, the contour aware convolutional neural network introduced by Chen *et al.* (2016) for gland segmentation aims at incorporating the observation that epithelial cell nuclei provide good boundary cues for splitting clustered objects. This model is trained on the original annotations of the gland objects from pathologists as well as on the automatically extracted boundaries with the purpose of learning explicit representations of the object contours. The two latter examples particularly illustrate how solutions can be designed to enforce machine learning and human cognitive capabilities to complement each other.

4.5.3 Machine Learning Teaming

Kontschieder *et al.* (2015) introduced the so-called Deep Neural Decision Forests. The authors propose a probabilistic routing alternative to the standard binary and deterministic routing strategy, which makes the optimization of a global cost function through backpropagation possible. Embedding the forest decision functions with the nodes of the fully connected output layer of a convolutional neural network enables representation learning. Deep neural decision forests have been shown to further improve the accuracies obtained with the prevailing random forests and deep learning approaches, thereby illustrating how teaming conceptually different machine learning approaches can lead to more effective systems.

4.5.4 Conclusion

As artificial intelligence is getting more and more mature, it is crucial to leverage the abilities of modern machine learning approaches to solve complex computer vision related problems. This is particularly true as some current limitations of deep learning approaches such computational efficiency on whole-slide histopathology images are about to be solved using, for instance, hierarchical strategies. It is, however, equally important to develop machine learning systems that use the best data-driven and knowledge-driven information for each specific application and not to take existing systems as out-of-the-box solutions only.

References

Brieu, N. and Schmidt, G. (2017). Learning size adaptive local maxima selection for robust nuclei detection in histopathology images. In *2017 IEEE 14th International Symposium on Biomedical Imaging (ISBI, 2017)*, Melbourne, VIC, pp. 937–941.

Brieu, N., Pauly, O., Zimmermann, J., Binnig, G., and Schmidt, G. (2016). Slide-specific models for segmentation of differently stained digital histopathology whole slide images. In *Proceedings of SPIE 9784, Medical Imaging 2016: Image Processing, 978410* (21 March 2016). doi: 10.1117/12.2208620.

Chen, H., Qi, X., Yu, L., and Heng, P.-A. (2016). DCAN: Deep contour-aware networks for accurate gland segmentation. In *2016 IEEE Conference on Computer Vision and Pattern Recognition (CVPR)*, Las Vegas, NV, pp. 2487–2496.

Cireşan, D. C., Giusti, A., Gambardella, L. M., and Schmidhuber, J. (2013). Mitosis detection in breast cancer histology images with deep neural networks. In *Medical Image Computing and Computer-Assisted Intervention—MICCAI 2013* (Springer, Berlin, Heidelberg), pp. 411–418.

Criminisi, A., Shotton, J., and Bucciarelli, S. (2009). Decision forests with long-range spatial context for organ localization in CT volumes. In *MICCAI Workshop on Probabilistic Models for Medical Image Analysis (MICCAI-PMMIA)*, pp. 69–80.

Kainz, P., Urschler, M., Schulter, S., Wohlhart, P., and Lepetit, V. (2015). You should use regression to detect cells. In *Medical Image Computing and Computer-Assisted Intervention—MICCAI 2015* (Springer, Cham), pp. 276–283.

Kontschieder, P., Fiterau, M., Criminisi, A., and Rota Bulo, S. (2015). Deep neural decision forests. In *Proceedings of the IEEE International Conference on Computer Vision*, pp. 1467–1475.

Krizhevsky, A., Sutskever, I., and Hinton, G. E. (2012). ImageNet classification with deep convolutional neural networks. In *Advances in Neural Information Processing Systems 25*, F. Pereira, C. J. C. Burges, L. Bottou, and K. Q. Weinberger, eds. (Curran Associates, Inc.), pp. 1097–1105.

Lecun, Y., Bottou, L., Bengio, Y., and Haffner, P. (1998). Gradient-based learning applied to document recognition. In *Proceedings of the IEEE*, **86**(11), pp. 2278–2324.

Loy, G. and Zelinsky, A. (2002). A fast radial symmetry transform for detecting points of interest. In *Computer Vision—ECCV 2002* (Springer, Berlin, Heidelberg), pp. 358–368.

Minsky, M. and Papert, S. (1972). *Perceptrons: An Introduction to Computational Geometry* (Cambridge MA, USA: The MIT Press).

Peter, L., Pauly, O., Chatelain, P., Mateus, D., and Navab, N. (2015). Scale-adaptive forest training via an efficient feature sampling scheme. In *Medical Image Computing and Computer-Assisted Intervention—MICCAI 2015* (Springer, Cham), pp. 637–644.

Rosenblatt, F. (1958). The perceptron: A probabilistic model for information storage and organization in the brain. *Psychol. Rev.*, **65**, 386–408.

Rumelhart, D. E., Hinton, G. E., and Williams, R. J. (1986). Learning representations by back-propagating errors. *Nature*, **323**, 533–536.

Szegedy, C., Liu, W., Jia, Y., Sermanet, P., Reed, S., Anguelov, D., Erhan, D., Vanhoucke, V., and Rabinovich, A. (2015). Going deeper with convolutions. In *2015 IEEE Conference on Computer Vision and Pattern Recognition (CVPR)*, Boston, MA, pp. 1–9.

Vinod, N. and Hinton, G. E. (2010). Rectified linear units improve restricted boltzmann machines vinod nair. In *Proceedings of the 27th International Conference on Machine Learning (ICML-10)*, pp. 807–814.

Wang, D., Khosla, A., Gargeya, R., Irshad, H., and Beck, A. H. (2016). Deep learning for identifying metastatic breast cancer. *CoRR* abs/1606.05718 (2016): n. pag.

Chapter 5

Image-Based Data Mining

Ralf Schönmeyer, Arno Schäpe, and Günter Schmidt
Definiens AG, Bernhard-Wicki-Strasse 5, 80636 Munich, Germany
rschoenmeyer@definiens.com

The objective of image mining in the context of tissue phenomics is the discovery of novel phenes with biomedical relevance. To achieve that goal, data mining techniques are applied to discover patterns in image analysis results, which correlate best with associated experimental result data. These patterns may comprise spatial relationships of cell population, or geometric properties of tissue architectures. The result data may encompass various levels of data acquired in translational sciences, but most importantly treatment response groups in clinical trials. One particular commonality with genomics is the curse of dimensionality: The number of potentially relevant phenes is much larger than the number of individuals in the study. In this chapter, we discuss the image mining methodology, from spatial analysis of millions of cells in virtually multiplexed studies using cell density heatmaps, to correlating potentially relevant patterns with clinical observations by optimizing predictive values and survival time prediction.

Tissue Phenomics: Profiling Cancer Patients for Treatment Decisions
Edited by Gerd Binnig, Ralf Huss, and Günter Schmidt
Copyright © 2018 Pan Stanford Publishing Pte. Ltd.
ISBN 978-981-4774-88-8 (Hardcover), 978-1-351-13427-9 (eBook)
www.panstanford.com

5.1 Introduction

Image mining, in general, describes the process of generating knowledge from information implicitly available in images. When focusing on applications in the field of histopathology, image mining of collections of digital whole tissue slides, generated from human patient samples with known disease progression, enables the extraction of diagnostic knowledge. Since digital pathology images are huge, comprising gigapixels of unstructured information, they have to be processed using image analysis methods to become mineable. The initial structuring of image pixels to biologically meaningful image objects, such as cell compartments, cells, and tissue architectures, is a prerequisite for the search for diagnostically relevant information.

Beyond data mining methods that operate on table data, image mining sets a strong focus on the geometry and topology of the extracted image objects. This imposes a challenge for the data mining process, because the number of potentially relevant patterns becomes virtually infinite due to the unlimited number of spatial grouping combinations to consider. Therefore, image mining in biomedicine has to operate within certain boundary conditions imposed by existing knowledge. The knowledge may comprise spatial parameters such as the effective range of cell-to-cell interactions, or which cell types are potentially involved in disease progression, or the potential mechanisms of action of a therapy used to treat the patient.

Mathematical methods involved in the data mining process comprise feature selection, supervised and unsupervised cluster analysis, and correlation analysis. Since all methods are frequently used together, their interdependence may require a nonlinear optimization method to generate the most accurate result. Another relevant class of methods addresses the problem of overfitting: Since the number of samples is much lower than the number of available image features, a strong emphasis has to be put on cross-validation workflows, together with stability analysis of the discovered diagnostic algorithms. Stability can be measured with respect to variations in image content, variations of parameters derived in the data mining process, or variations in the clinical parameters of disease progression.

Introduction | 103

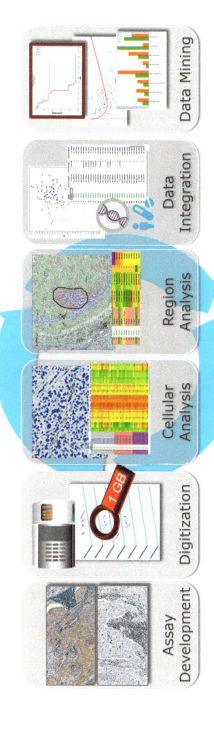

Figure 5.1 Illustration of the tissue phenomics workflow from the assay development, to tissue slide digitization to the interactive analytics. This chapter covers the aggregation of features in regions of interest, the integration of clinical data, and the analytics to generate algorithms for personalized immuno-oncology.

This contribution explains the workflow from basic image analysis results to a diagnostic algorithm in detail, describing various options with their respective advantages. An overview is given in Fig. 5.1. It sets a strong focus on image mining of immunohistochemically stained tissue sections, but the underlying principles are applicable to other modalities such as RNA or *in situ* hybridization–based tissue-staining methods, or even to radiology such as computed tomography (CT) and magnetic resonance imaging (MRI).

5.2 Generating High-Level Features for Patient Diagnosis

Image mining requires the aggregation of all measurements for a single patient into a single feature vector. This vector comprises various numerical and categorical entries, which provide the input for the subsequent data mining methods. In the image mining of histopathology images, the image content is frequently aggregated using statistical operations in specific regions of interest (ROIs). Common aggregation methods are minimum, mean, quantiles, or maximum value of the measurements within the ROI. At first, we discuss ROIs, which are generated by expert pathologists, and next regions, which are detected automatically using high-level image analysis.

5.2.1 Quantification of Regions of Interest

The human visual perception is well trained to detect a variety of tissue types in pathology images, particularly in hematoxylin and eosin (H&E)-stained images, depicting the morphology of cells in great details. By utilizing this ability, the pathologist may delineate tumor regions using a digital pathology workstation, enabling the software to determine cell counts within that region. Since the area of the delineated region is known, the density of the cells corresponding to a given class can be easily calculated. Relevant classes are marker-positive cells such as KI67(+) cells depicting proliferating cells, or CD8(+) cells depicting cytotoxic T cells. In case the image analysis algorithm is capable of detecting marker-negative tumor cells, the percentage of marker-positive cells within the annotated region can be measured. Application examples are the KI67 proliferation index or the HER2 score within the annotated ROI.

Beyond measurements with respect to a single cell type, average spatial relationships within the annotated ROI can be determined. Examples are the density of lymphocytes that are located within 100 μm from cancer cells as determined by the image analysis software using H&E-stained tissue section, or the density of CD8(+) cells in the vicinity of FOXP3(+) cells using double-stain immunohistochemistry (IHC).

Since the number of immunohistochemical markers in each tissue section is limited to three or four in bright-field microscopy, the method of serial sectioning has been used to combine several, sequentially cut tissue sections to one *information block*. The serial sections may be aligned using automated image co-registration methods (Yigitsoy *et al.*, 2017), which enables the computation of co-occurrences of various cell types identified in the individual sections. The accuracy of this method is restricted by the rate of change of the tissue from one section to the other, and by the accuracy of the co-registration algorithm. Although it has been shown that a hierarchical patch-based co-registration may produce very high accuracy, the limitations arise from tissue processing artifacts such as folds or tears.

Image co-registration helps twofold: first, the pathologist annotation in H&E can be automatically transferred to the corresponding IHC sections, and second, the measurements in each ROI of each tissue section may be combined using arithmetical operations (Schönmeyer *et al.*, 2015). Therefore, the joint probability of occurrence of two cell types may be expressed as the geometric mean of the corresponding average cell densities, or two potential therapeutic targets as the maximum of the respective cell densities. As already mentioned in the introduction, the combinatorial multiplicity requires the restriction to meaningful expressions, such as the geometric mean to PD-L1(+) and CD8(+) for an anti-PD-L1 checkpoint inhibitor therapy.

5.2.2 Heatmaps

As already mentioned, full-scale digital pathology images are usually too big to be processed completely at once in the working memory of personal computers. Heatmaps are images with demagnified dimensions of the original images that contain aggregated information for the full image. They computationally can be handled conveniently, because their extent usually is in the range of a few hundred to a few thousand pixels, and not like in full-resolution whole-slide images (WSIs) with several hundred thousand pixels for length and width.

5.2.2.1 Contents of heatmaps

Each pixel of a heatmap represents a tile of its corresponding region in the original image and carries statistical data spatially related to that tile. These statistical data are aggregated from detailed image analysis results in or near each tile and typically exhibit densities from objects of interest. For example, a single-layer heatmap represents the density distribution of a certain cell type like KI67(+) tumor cells (see Fig. 5.2a,b). Usually, heatmaps consist of many such layers with statistical data from a variety of such measurements. This comprises density measures of objects of interest but also more abstract information such as mean distances between two types of different objects of interest. This way heatmaps contain non-trivial aggregated information of the full image and allow for efficient further processing. This also enables optimization loops to adjust model parameters, which computationally would by far be too expensive when performed on full-scale images (Schoenmeyer *et al.*, 2014).

Heatmaps require image analysis results, for example, basic data like classified cells/nuclei in immunohistochemically stained WSIs. Densities or amounts of certain cell types are computed on such basic data and are stored as intensity values in distinct heatmap layers.

Information derived from high-level areas such as ROIs from annotations can contribute to heatmap layers as well. In general, the resolution of heatmaps should be chosen in a way that the structures of interest in an image keep their characteristic appearance for reasonable follow-up processing. When dealing with co-registered data, the registration quality of serial sections in full resolution should be within the tile size of corresponding heatmap pixels.

Generating High-Level Features for Patient Diagnosis | 107

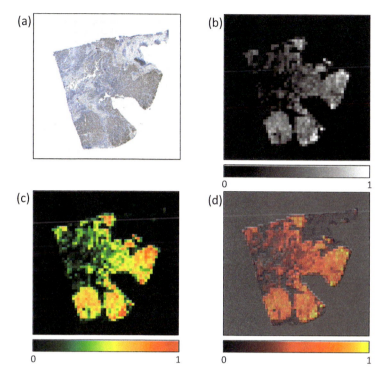

Figure 5.2 Visualization of tumor cell proliferation density: (a) overview of a WSI with KI67-stained breast tissue, (b) gray scale heatmap of KI67(+) tumor cell density, (c) green–yellow–red false-color representation of density, (d) red–yellow false-color representation of density as overlay to WSI overview.

5.2.2.2 Visualization of heatmaps

To visualize heatmaps, the most simple and obvious way is to display the intensities as a grayscale map (Fig. 5.2b). Intensities are normalized to fully or partly utilize the available grayscale ranges of a display, for example, 0 to 255 intensity steps in 8-bit image representation or 0 to 65535 for 16-bit. Furthermore, intensities can be displayed as false-color representations and some common ones are shown in Fig. 5.2c,d.

To visualize the contents of three heatmap layers in one map, an established approach is to show them as the three channels in RGB color space (Fig. 5.3). When heatmap layers carry intensities of different magnitude or certain heatmap layers should be shown more/less prominent, weights can be applied to account for that.

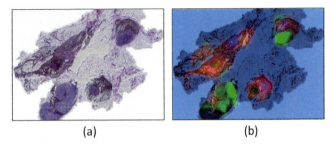

Figure 5.3 Visualization of three heatmap layers as RGB overlay: (a) overview of a WSI with H&E-stained skin tissue; (b) density of melanocytes (red channel), CD3(+) cells (green channel), and FOXP3(+) cells (blue channel) as overlay to WSI overview. The densities originate from co-registered data and are slightly smoothed for visualization to accommodate uncertainties from the limited co-registration accuracy.

Smoothing of heatmaps often helps to make them look more appealing. Especially when heatmaps have to be displayed magnified, for example, as overlay to the original image data, smoothing hides the fact that their data consist of comparably coarse pixels.

Finally, and since heatmaps can be used as any other image, there are no limits for data visualization. As an example, Fig. 5.4 shows the values of a heatmap as three-dimensional height profile textured with the original WSI data.

Figure 5.4 Heatmap data as profile: In a three-dimensional visualization of a WSI with KI67 tissue, the density of KI67(+) tumor cells is used to define the height in the profile.

5.2.2.3 Heatmaps with multiplexed data

Working with multiplexed data means that different sources for quantitative measurements of the same location (here, the heatmap's pixels) are available for common computations.

WSIs of consecutive tissue slices can be virtually aligned in a way that corresponding regions match to each other in a common coordinate system such as shown in Fig. 5.5. This is called co-registration, and it is performed using automated landmark detection (Yigitsoy *et al.*, 2017).

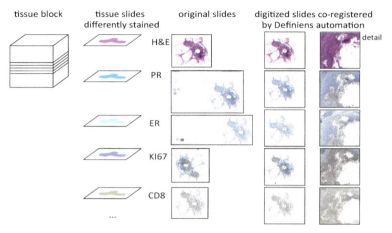

Figure 5.5 From a tissue block, serial sections are taken and differently stained. Since manual steps are involved, resulting tissue slices have varying positions and orientations on their glass slides and subsequent scanned images. With the help of automatically detected landmarks, the slides are aligned in a way that corresponding regions overlap as much as possible and computations in a common coordinate system can be performed.

When the slices are differently stained, each WSI carries specific information of the stain and automated image analysis makes this information available for computations. Using co-registration, these data are fused into one common coordinate system and allow for generating heatmap layers with multiplexed data of different stains (Schönmeyer *et al.*, 2015). Additionally, relations between statistics of differently stained sections can be processed and stored in heatmaps.

To avoid introducing errors by co-registration, an alternative way to obtain multiplexed data for heatmaps is the evaluation of one single tissue slice that carries information of several markers. With bright-field microscopy, usually no more than two to three markers are available, and with fluorescence techniques, more markers can be encoded (Stack *et al.*, 2014; Feng *et al.*, 2016).

Within the tissue phenomics framework, working with multiplexed data is supported by *virtually multiplexed Definiens Result Containers* (vmDRCs). In contrast to standard DRCs, which in general store image analysis results from a single WSI only, vmDRCs integrate image analysis results of several co-registered WSIs. DRCs are HDF5 compatible databases, capable of storing image and structured tabular data. Each database is stored in a single file, enabling easy sharing and backup. Heatmaps based on multiplexed data can directly be generated from vmDRCs, and there is no need to deal with co-registration furthermore because this happened already while preprocessing.

5.2.2.4 Objects in heatmaps

Since heatmaps consist of pixels, there is no restriction to use them as input for subsequent image analysis. This includes generation and manipulation of objects as shown in Chapter 3, *Cognition Network Technology*.

A typical use case is *hotspot detection* to define ROIs. Therefore, a layer in a heatmap is composed and a threshold is applied to generate objects that exceed the threshold. Figure 5.6 gives an example where regions with a high density of KI67(+) tumor cells define ROIs in WSIs. These ROIs are then regarded as hotspot objects, which are complemented by other heatmap layers. This way, objects with rich feature vectors form the base for follow-up data mining procedures.

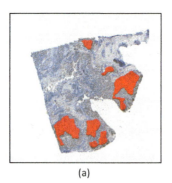 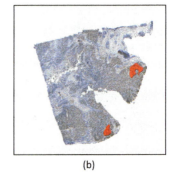

(a)　　　　　　　　　　　　(b)

Figure 5.6 Hotspot objects in heatmaps: Regions of KI67(+) tumor cell density that exceed a threshold are regarded as hotspot objects for further data evaluation. (a) Objects generated with a threshold of 150 for 8-bit normalized density and (b) a threshold of 180.

Generating High-Level Features for Patient Diagnosis | 111

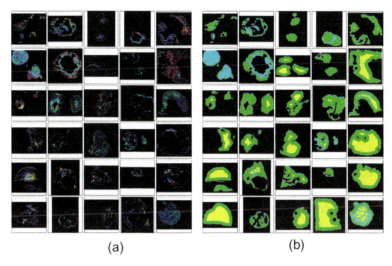

(a) (b)

Figure 5.7 Unsupervised clustering for object creation: (a) RGB visualization of heatmaps for density of melanocytes (red), CD3-positive cells (green) and FOXP3-positive cells (blue) for different co-registered tissue samples as depicted in Fig. 5.2. (b) Based on these layer values, an unsupervised clustering procedure identifies similar regions across different cases. The number of class labels (clusters)—here, three (blue, green, yellow)—is determined by optimizing the Rand index, quantifying similarity of class labels in training and test data.

Another approach for defining regions/objects of interest in heatmaps uses unsupervised clustering (see Fig. 5.7) (Rand, 1971). Therefore, the pixels of heatmaps from a cohort are characterized by the heatmaps' layer values and undergo an unsupervised clustering method like *k-means*. For a given number of classes *k*, the heatmap pixels are partitioned into *k* regions. Each class represents a specifically characterized type of region across the cohort's heatmaps. Therefore, all regions of the same class subsume similar regions in each WSI. The resulting objects again are quantitatively enriched by heatmap layer values as feature vectors and are suited to constitute the base for further data mining.

5.2.3 Algebraic Feature Composition

As we have seen in the previous sections, automated image analysis of ROIs or feature heatmaps generates multiple features for each patient. Sometimes it is possible to correlate one of those features with patient metadata, such as therapy success. One example is the proliferation index for breast cancer prognosis, which is based on the percentage of KI67(+) tumor cells in selected fields of view. Another example is the HerceptTest™ (Dako) algorithm, which uses the percentage of HER2(+) tumor cells to predict Herceptin® therapy success. In most cases, however, it is required to compose a score from various measurements to deliver predictive power. The Immunoscore®, for instance, combines CD8(+) and CD3(+) cell density measurements in the tumor center and in the invasive margin to a final score predictive of colon cancer survival.

Algebraic feature composition automatically generates scoring algorithms by combining features using algebraic or logical expressions. Algebraic combination comprises elementary operations such as *addition, multiplication, division*, whereas logical operations comprise *and, or,* and *exclusive-or*. To generate all bivariate combinations, a brute force approach combines any two features of the available feature space with any operation to generate a potentially predictive score. For N features, the bivariate combination approach produces $N \times (N-1)/2$ potential scores for each algebraic expression, whereas for each score, the patient subgroup of interest may be either higher than a given or computed threshold, or lower.

In the same way, all triple and higher order combinations may be generated. This approach provides a huge derivative feature space, which is computationally challenging and which increases the risk of overfitting. Both issues are discussed more generally in the section on feature space reduction.

To avoid the comprehensive combinatorial screening of all potential feature combinations, genetic programming has been proposed. Krawiec (2002) discusses an evolutionary framework that combines feature generation and selection. The selection process is driven by optimizing the prediction accuracy in a cross-validation setting.

Another implementation of feature generation comprises the variation of a parameter that is associated with a feature. For example, the classification of CD8(+) cells in the tumor microenvironment depends on the definition of the tumor border size, which can be chosen anywhere from 50 μm to 500 μm, and depending upon the actual chosen size, the density of CD8(+) cells within the tumor microenvironment varies. The introduction of continuous parameters in the feature generation process introduces a feature space with infinite dimensionality.

5.2.4 Aggregation of Multiple Tissue Pieces to Single Patient Descriptors

Considering the tumor heterogeneity and the tiny volume of tissue in a section on a glass slide, it is good practice to analyze several sections of a tissue block to obtain a statistically meaningful measurement. For tissue microarrays, we consider at least three cores to be acceptable, whereas in the diagnosis of prostate cancer, at least 10 biopsies with three sections, each at a different level, are examined.

In the light of a comprehensive phene discovery approach, we test all possible selection and combination rules of measurements (features) on these tissue pieces. The most straightforward combination rules are as follows:

1. Virtually combining all tissue pieces into a single piece and measuring therein the feature, such as the percentage of marker-positive cells.
2. Computing the area-weighted average (standard deviation, quantile) of the feature measured in each tissue piece.
3. Computing the maximal (minimal) feature value of all tissue pieces.

Although the first approach may deliver the most accurate value for an average measurement, the method closest to a pathologist examination is the third, in which only *relevant* tissue pieces are considered. Since *a priori* the *best* method in terms of predictive power is not known, several data aggregation strategies may be included in parallel, which may be a natural extension of the algebraic feature composition concept.

5.2.5 Integration of Clinical Data and Other Omics Data

The phene discovery process in tissue phenomics aims to find optimal predictors for the patient's disease progression in the context of a therapy decision. The therapy success can be quantified by observing the progression-free survival (PFS), overall survival (OS), objective response rate (ORR), adverse events, or quality of life. In clinical trials for cancer therapies, the ORR is estimated according to the RECIST (Response Evaluation Criteria in Solid Tumors) criterion using CT imaging, which assigns the change in the tumor size to the corresponding response category, such as progressive disease (PD), stable disease (SD), partial response (PR), or complete response (CR). PR and CR are considered therapy responders. In clinical trial studies with OS or PFS as the clinical endpoint, the time from initial diagnosis to event (*e.g.*, death, disease progression) or last examination without event (left-censored) is recorded for each patient. These recordings are the basis of Kaplan–Meier survival analysis, which is one of the central modules of any clinical tissue phenomics project.

Other variables associated with a patient sample are the origin of the tissue, its pathological grade such as mitotic count, or IHC scores such as estrogen receptor (ER), progesterone receptor (PR), or HER2 status. Since cancer is a genetic disease, various mutation status such as EGFR, BRAC12, and BRAF may further enrich the patient profile. And, most obviously, the patient's demography, such as histology, age, gender, tumor stage, line of therapy, liver metastasis, and smoking habits, contributes with additional predictive value to the tissue-based parameters.

In a clinical research setting, the tissue may be analyzed using next-generation sequencing (NGS), gene expression (mRNA-based), and protein expression (antibody-based) arrays, potentially providing thousands of additional data points per patient. When looking toward larger spatial scales, radiomics-based features such as volume, shape, and texture of tumors as seen by CT or MRI may contribute substantially to the high dimensionality of the available feature space (Fig. 5.8).

As an interesting side note, image analysis of (IHC or H&E) stained tissue sections may provide normalization and de-convolution information for data acquired by non-spatially resolved modalities such as gene expression arrays. In such an approach, the gene signature of a cell population may be determined by identifying the best correlation of the gene expression measurements with the tissue-based detection of a cell population, such as lymphocytes in H&E (Hackl *et al.*, 2016).

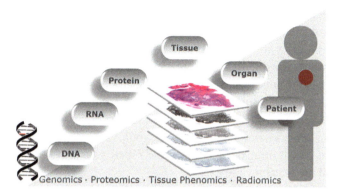

Figure 5.8 Tissue phenomics bridges the biomedical sciences from genomics to patient-level studies, while providing the highest spatial resolution for detailed quantification of cellular organization and function. Integrative tissue phenomics enriches information from digital histopathology with other modalities, such as genomics and radiomics.

Since we may combine all pre-treatment clinical parameters for a multi-parametric phene discovery process, we face the challenge that with a limited study size in the order of 50 to 500 patients, an almost infinite number of data points per patient may provide potentially predictive value just by chance. To overcome this challenge, advanced feature selection methods are required.

5.3 Performance Metrics

Before we proceed toward feature selection methods, we should discuss the metrics by which we evaluate the predictive quality of a phene with respect to disease progression. In its simplest instantiation, a phene consists of a feature, and a logical threshold-based (cut-point) decision on how to evaluate the feature value to categorize a patient into a disease progression group. In the context of companion diagnostics for personalized healthcare, in which the phene-positive [phene(+)] patients respond best to therapy (CR or PR groups), the following performance indicators are considered to be particularly important (Fletcher and Fletcher, 2005).

The *prevalence* measures the proportion of patients that are phene(+). The computation of the prevalence does not require the knowledge of disease progression; it is an inherent property of the phene and the patient cohort under consideration. The higher the prevalence, the more patients will receive the therapy, which is an important factor for the economic success of a therapeutic option. In the case of missing response and survival data, the prevalence may be the only way to select a cut-point for a feature. Since the distribution statistics of feature values across a cohort frequently does not provide a hint on a *natural* cut-point selection, such as with bimodal distributions, the cut-point is set to a quantile (*e.g.*, 50%, 66%) of the feature values.

The *positive predictive value* (PPV) is defined as the proportion of patients who respond to therapy within the subpopulation of phene(+) patients. The higher the PPV, the more predictive the phene.

The *negative predictive value* (NPV) is defined as the proportion of patients who do not respond to therapy in the group of phene(−) patients. This value should be very large, in many applications >90%, because we do not want to refrain patients from therapy due to the phenomic selection.

The Kaplan–Meier *log-rank test* returns a *p-value* of the PFS (pPFS) or OS (pOS) when comparing two patient groups of a clinical study, such as phene(+) patients versus standard of care (SoC), or phene(+) versus phene(−) patients. *p*-Values smaller than 0.05 are considered significant.

The *accuracy* measures the prediction quality in terms of the ratio of correctly predicted response category (PD/SD versus PR/CR). This measurement is independent of the PFS time and should, therefore, be considered supplemental.

The variety of performance metrics renders the optimization of a phene as a multi-objective problem. One heuristic approach to solve that problem is to select that performance metric for optimization, which is associated with a primary endpoint in the clinical trial under consideration. All other performance metrics are allowed to float in a predefined range, such as $NPV > 0.9$, $0.2 < PPV < 0.4$, and $prevalence > 0.5$. In Fig. 5.9, we show how the selection of a cut-point influences the phene performance with the help of an example. Another approach would optimize the PPV by enforcing the log-rank test to be significant.

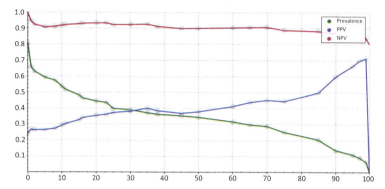

Figure 5.9 Illustration of the dependence of prevalence and predictive values (PPV and NPV) for a chosen cut-point (x-axis). A "window of opportunity" seems to be open for a cut-point in the range of 20–50 where both PPV and prevalence are robust for feature value variations. Given that range, the tissue phenomics optimization searches for the cut-point with the minimal pPFS or pOS.

5.4 Feature Selection Methods

This section discusses briefly various options to select the most relevant feature, which is the main component of a predictive phene. To recapitulate, a feature is an element of a vector of numerical values describing the status of the patient prior to therapy, where most features originate from tissue-based measurements and from subsequent algebraic multivariate composition.

5.4.1 Unsupervised Methods

Since many features show a strong correlation with other features, one strategy for feature selection is to perform an unsupervised cluster analysis and pursue with those features close to the cluster centers. In such an analysis, the clusters are defined by the distance of feature i to feature j, where the distance is measured, for example, as the weighted Euclidean distance of the vector of all feature i values of all patients to all feature j values of all patients.

To enable feature comparison, in particular for unsupervised clustering, we normalize the feature values. Several normalization methods are available (Aksoy and Haralick, 2001); the most important are as follows:

- In quantile-based normalization, a lower (q_l) and a higher (q_h) quantile of the feature values f_i are computed. The normalized feature value is determined by $f_i' = (f_i - q_l)/(q_h - q_l)$. A simplified application of this method is the min–max normalization, in which the lower 0% quantile and the upper 100% quantile are chosen. This ensures that the resulting feature values are within [0...1]. To avoid the immoderate impact of outliers to this simple normalization, a robust outlier detection method must be implemented before min–max normalization is applied.
- For feature values with a Gaussian distribution, which is actually rather the exception than the norm, we may normalize for mean (μ) and standard deviation (σ): $f_i' = (f_i - \mu)/\sigma$. More general approaches use the density of the distribution as the basis for normalization. The histogram equalization known from image processing is one specific application using the cumulative distribution function of pixel values in an image (Acharya and Ray, 2005).

- If no assumption on the statistical distribution can be made, then a rank normalization may be beneficial, in which the feature values are replaced by their ranks in the sorted feature list divided by the number of patients. The resulting feature values are ensured to be in the range [0...1] (Tsodikov *et al.*, 2002).

Dependent on the specific implementation, the number of clusters (*e.g.*, in centroid- and hierarchical-based clustering, see Fig. 5.10) or the homogeneity of clusters (*e.g.*, in connectivity clustering) must be determined beforehand, or constitute another parameter in the overall optimization loop. To evaluate the quality of the cluster analysis, the median-adjusted Rand index and its confidence interval should be computed using multiple, randomly chosen data subsets (Rand, 1971).

5.4.2   Hypothesis-Driven Methods

In hypothesis-driven methods, we restrict the available feature space to known drivers for disease progression. Frequently such information is extracted from scientific literature using advanced text mining methods such as *GoPubMed* (Delfs *et al.*, 2004).

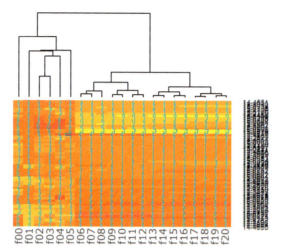

Figure 5.10 Example for feature selection by unsupervised hierarchical clustering. Each row in the heatmap represents a patient, each column a feature (feature 00 to feature 20). The heatmap colors correspond to normalized feature values. The dendrogram on the top visualizes the result of an unsupervised hierarchical clustering. If we would restrict the supervised feature selection to two relevant features, one option is to choose the centroids of the topmost clusters, which may be feature 02 and feature 13.

Another source for hypotheses is well-structured databases on protein signaling networks, such as Kyoto Encyclopedia of Genes and Genomes (KEGG) (Kanehisa *et al.*, 2017). These databases are curated and comprise the best common understanding about biology. However, we are not aware of a database of cell-to-cell signaling networks, which are most important in understanding the immune contexture of tumors. In addition to that, the mechanistic understanding of cancer progression should involve the knowledge of the cancer's mutational status and which pathways are eventually changed due to the genetic alteration. This limits the hypothesis-driven approach to known cancer signaling pathways, for example, those involving KRAS or BRAF gene products.

The hypothesis-driven methods can confirm or reject a known hypothesis in the context of the given tissue phenomics study. But if we solely rely on known hypotheses, we will not discover novel biology. The ideal method, therefore, combines data-driven with hypothesis-driven approaches.

5.4.3 Univariate, Data-Driven Feature Selection

As we discussed the generation of multivariate features in the algebraic feature composition section, we restrict ourselves in this paragraph to the univariate selection of a feature using information on disease progression. The primary challenge for feature selection is the low number of patients (in the order of hundreds) compared to the high number of available features and prediction models (in the order of thousands). Therefore, we need methods that estimate the predictive power of any model in unseen data. One well-established method to achieve that goal is cross-validation (Bornelöv and Komorowski, 2016). We extended the method by Monte Carlo *feature validation* of the available feature space, and the Monte Carlo *feature selection* of the best model by bootstrapping (Fig. 5.11).

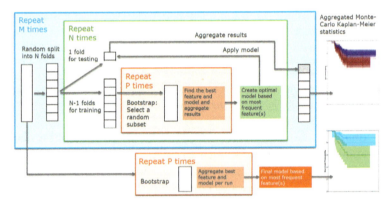

Figure 5.11 Monte Carlo cross-validation for feature validation (top) and selection (bottom).

For *feature validation*, the complete dataset is randomly divided into N parts (folds), for which $N-1$ folds are used to train a model M_{N-1} (feature and decision rule). The model M_{N-1} is chosen by bootstrapping, using a random subset within the $N-1$ folds and selecting the most frequently chosen bootstrap model. Each of these bootstrap models is optimized for prediction performance in terms of log-rank test p-values, predictive values, and prevalence. The model M_{N-1} is applied to the remaining 1 fold. After N iterations, all N folds have been used for prediction once using the found optimal models, so that the cross-validated aggregated performance for the complete dataset can be determined and recorded. This process is repeated several times to generate statistics on the robustness of the selected models and within the context of available features as shown in the Kaplan–Meier curves in Fig. 5.11 top-right. Each curve pair represents the result of one outer loop iteration for random fold generation and gives an estimate of the expected generalization performance of the features selected in the inner-most loop when applied to classify new unseen data.

For *feature selection,* we restrict the process to bootstrapping (P times random subset selection), in which we choose the model that was most frequently selected in the P loops model selection process, where only models are considered, which fulfill given restrictions on the performance values such as Kaplan–Meier log-rank p-value. The resulting model's performance may be visualized by a single Kaplan–Meier plot (Fig. 5.11 bottom-right).

Using the above-described feature validation and selection, we obtain information on which feature may deliver the most robust predictive value. However, we still need to perform a permutation test to compute the probabilities of obtaining a similar prediction performance using randomly chosen clinical outcome information (Radivojac *et al.*, 2004). Since the statistical distribution of the clinical endpoint data should not be different from the original data, the permutation test shuffles the observed outcome data randomly for each run. Using the shuffled endpoint data, the steps described above are repeated, so that we compute in each permutation step the best model and its performance. The probability of having selected a model by chance is then estimated as the percentage of permutation test runs, which deliver a better predictive model.

5.5 Tissue Phenomics Loop

The mathematics of tissue phenomics may be considered an optimization problem. All steps, from assay and patient cohort definition to image analysis; from aggregation of image analysis results to ROIs; and from algebraic feature composition and subsequent feature selection to the correlation to disease progression require defining of continuous and categorical parameters that finally determine the quality of the most predictive phene. The straightforward solution to this optimization problem by using one algorithm is not feasible due to the huge dimensionality of the feature space. Therefore, we decided to perform multiple optimizations in a heuristic approach, optimizing the parameters within each step, and coupling the single-step optimizations by several meta-optimization loops (see Fig. 5.12).

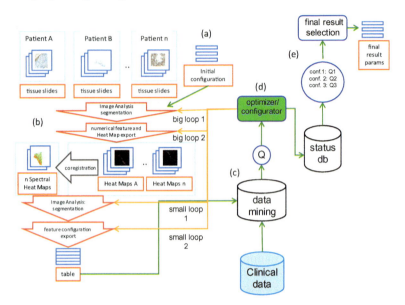

Figure 5.12 Exemplary workflow of a tissue phenomics loop to optimize for clinical outcome values using heatmaps: (a) WSIs from patients undergo an image analysis procedure governed by an initial configuration. (b) Numerical features and heatmaps are generated. The heatmaps are multiplexed using co-registration, and an image analysis produces a table with features describing each patient. (c) The results are stored in a database used for data mining to model endpoints from clinical data. (d) Based on this, a quality Q is assessed and an optimizer/configurator module decides for the next processing steps: A new configuration is produced (and stored in a status database) and defines at which entry point—and thus level of complexity—the loop is continued: *small loops* 1 or 2 vary parameters for the analysis based on heatmaps, where the latter keeps the current heatmaps as is and only varies the configuration for feature export to the database. In contrast, *big loop* 1 is computationally very expensive because it involves a reprocessing of the image analysis segmentation for full WSIs. *Big loop* 2 is less extensive and reconfigures the feature and heatmap production based on existing image segments in WSIs only. In either case, new result data are produced, which enriches the data mining database. When a certain quality Q has been reached, the loop is stopped and a final quality assessment and result configuration selection is performed. The final set of result parameters constitutes the configuration that evaluates WSIs best for clinical endpoint modeling.

In particular, this approach considers the computationally expensive optimization of the image analysis in the context of clinical data. An inner loop optimizes image analysis results toward ground truth data as provided by domain experts, and an outer loop optimizes some less defined parameters such as the decision rule on marker-positive or -negative cells to the disease progression information.

5.6 Tissue Phenomics Software

The overall tissue phenomics analysis chain consists of several steps, each bearing its own inherent complexity. Proper software solutions are a necessary prerequisite for successful application of the described approach. The width of the overall requirements as well as the depth of the requirements within each area represent a significant challenge to any software solution. Therefore, an overall tissue phenomics software solution can only be built by creating a system comprising tightly integrated specialized software components.

5.6.1 Image Analysis and Data Integration

The first challenge any software system for tissue phenomics needs to tackle is image analysis (see Chapters 2–4). The unstructured data represented by digital tissue slide images need to be converted into structured numerical or categorical data that can be used as input for data mining and optimization methods and which can be linked with numerical and categorical metadata from other sources. As described above, the metadata may be patient-centric clinical data or data from other modalities such as genomics or flow cytometry in a multi-omics approach.

5.6.2 General Architecture

In addition to the data integration requirements, a tissue phenomics software system aiming toward the deployment of the discovered phene into clinical route covers the aspects of regulatory compliance and user, data and workflow management (Fig. 5.13). The system integrates multiple stakeholders in this multidisciplinary process:

pathologists reviewing the slides, providing annotations and quality control assessments. Image analysis experts apply image analysis solutions tailored to specific questions and adopt to assays and quality and performance requirements. Data scientists perform data mining and signature optimization tasks, and project managers keep track of the project status and decision makers review the outcome of the process.

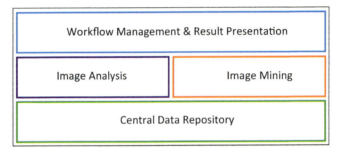

Figure 5.13 Schematic illustration of a tissue phenomics software infrastructure. The first layer (blue) focusses on end users that are interested in the progress and outcomes of tissue phenomics projects by offering a "portal" view on the data. Here the flow of the data within a project can be managed, and the status and available results are presented, quality control steps are performed, and results are presented. The second layer provides the analytical components. These are separated into an image analysis system (purple), which is optimized for converting digital tissue slides into a numerical and categorical representation and a data mining system (orange) that is optimized for the data integration and analysis tasks using computer readable numerical and categorical data from many different sources. The third layer (green) provides the central data repository for the entire system. All relevant data are stored and managed within this layer. This layer is especially important if tissue phenomics results will be used for decision making within regulated environments. In this case, it needs to enable the necessary level of regulatory compliance for the overall system.

5.6.3 Image Mining Software

To support all the concepts described in this chapter, a tissue phenomics software system must implement a very scalable and adaptable table model. The aggregation of data from the cellular tissue level into higher level features such as ROIs or up to the patient level across studies should work very efficiently on tables with large numbers of rows. For typical tissue phenomics projects, cellular tables are in the range of 10 million to 1 billion rows with 10 to 10,000 columns. The table model also needs to be able to handle a large number of columns, since the number of aggregated and combinatorial features can be very large, in particular when data from other sources such as NGS or RNA assays have been integrated.

The data table processing needs to be very performant to deliver instant feedback to the user. Even if an overall optimization flow using all the methods described in this chapter may not require real-time processing, it is mandatory for the development of new tissue phenomics methods to enable quick turnaround times for the basic operations like feature composition or aggregation. One of the most important operations is the classification of all cells in a clinical study with respect to their spatial and contextual relationships. The technical foundation to achieve that goal in real/near time is the use of randomized k-d-trees, which enable the fast search for nearest neighbors as well as range searches within a given radius (Muja and Lowe, 2014).

Finally, the visual representation of intermediate and end results in different representations is extremely important for the efficient development of tissue phenomics–based image mining methods as well as for the interpretation of potential findings (Schönmeyer *et al.*, 2014). Therefore, we recommend interactive software systems with charting that is linked with the tabular data as well as with the image data, enabling users to understand the findings and outcomes of the mining process (Figs. 5.14 and 5.15). These capabilities are also essential for the quality control steps throughout the process, which are a specific type of image mining with the purpose of finding errors or inconsistencies in the data.

5.6.4 Data and Workflow Management and Collaboration

The central data repository manages and stores data relevant to the tissue phenomics processes. Many system requirements have a high degree of overlap with capabilities of enterprise content management systems (ECMS); however, the analytical aspects of tissue slide images and the related data mining impose an additional set of special requirements.

126 | Image-Based Data Mining

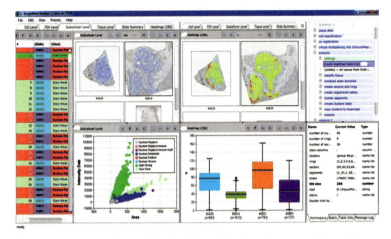

Figure 5.14 Screenshot of the Definiens big data image mining software. Left: Interactive large-scale table browser supporting 100s of millions of rows. Middle-top: interactive full-slide display of cells, regions, and heatmaps. Middle-bottom: interactive charting of small- and large-scale table data. All data viewing components are linked and enable interactive data exploration. Right-top: scripting environment supporting real-time interactive as well as fully automated analytics. Right-bottom: development and debugging support.

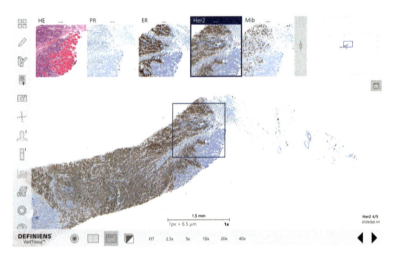

Figure 5.15 Exemplary screenshot of a web-based collaborative slide viewing, annotation editing, and result reviewing software portal, which can be deployed in cloud infrastructures. It comprises automated alignment of serial tissue sections to enable analysis of cell co-occurrences in the context of tumor heterogeneity, such as ER-positive and HER2-positive tumor regions.

Core tasks of the data repository are to store and organize all the data involved in the process and to track who performed which operation on which data at which point in time, together with providing access control, content-based searching capabilities, and workflow modeling. In particular, projects dealing with clinical trial data may fall under the US Food and Drug Administration (FDA) guidance on *Current Good Manufacturing Practice for Drugs*, which defines the minimum standards for data integrity and auditing. Other regulatory standards comprise electronic records and signatures (21 Code of Federal Regulations Part 11) and patient privacy (Health Insurance Portability and Accountability Act of 1996).

Specific to tissue phenomics are the requirements for appropriate storage and streaming of the large datasets involved, especially for tissue slide images and the related image analysis results data. As these datasets are in the range of several terabytes in operational use cases, specific measurements are necessary to support efficient processing and visualization of these datasets. This may include the use of solid-state drives or even random-access memory to store the most frequently accessed data and index structures and the use of distributed data centers to ensure highly-secure long-term storage.

As tissue phenomics projects also have a complex workflow and require collaboration between the many different stakeholders within the process, the use of browser-based technologies and cloud infrastructures to enable a geographically distributed ecosystem is essential for operational use, for example, to support multinational clinical trials or the collaboration of large pharma companies with their contract research organizations around the tissue phenomics datasets.

5.7 Discussion and Outlook

Image-based data mining is a complex process that transforms the terabytes of image-related information into a few bits of clinically relevant algorithms (phenes) capable of predicting the disease progression of a patient. The spatial characteristics of the underlying tissue image analysis results impose an additional challenge in terms of computational complexity. The infinite number of potential data aggregation mechanisms calls for unsupervised or hypothesis-driven methods to select the most relevant features, but in restricting ourselves to these approaches, little novel discoveries will be made. Therefore, we deeply believe in the power of big data and *top-down* approaches, in which novel insights are generated by unbiased data mining. To avoid overfitting models and conclusions that cannot be validated in prospective studies, the latest Monte Carlo validation methods are used to ensure robust phene selection. Interactive analytics link images to image analysis results and the results from phene discovery, enabling fast sanity checks using pathologist/oncologist domain expert knowledge.

The future will be even more dominated by unbiased machine learning approaches, such as deep learning. Currently, these methods are limited by the small number of patients, which prevents successful end-to-end learning, but the combination of auto-encoders and sparse-annotation algorithms will definitely impact the field of tissue phenomics. Machines have the power to *think* in multidimensional spaces, nonlinear feedback loops, and integrating knowledge of the disease progression of millions of patients. Therefore, it will be only a matter of time till humans are outperformed in providing the *best* diagnosis and therapy recommendation.

Acknowledgments

We thank Paolo Ascierto (Medical Oncology and Innovative Therapy at the National Tumor Institute Fondazione G. Pascale in Naples, Italy) for providing data and consultancy in the phene discovery study in melanoma (Figs. 5.3 and 5.7). Nathalie Harder, Carolina Vanegas, and Victor Matvienko (Definiens, Munich) contributed substantially to the cross-validation workflow and its implementation as described in Section 5.4.3 and Fig. 5.11.

References

Acharya, T. and Ray, A. K. (2005). *Image Processing: Principles and Applications* (Hoboken, N.J: John Wiley & Sons).

Aksoy, S. and Haralick, R. M. (2001). Feature normalization and likelihood-based similarity measures for image retrieval. *Pattern Recognit. Lett.*, **22**, 563–582.

Bornelöv, S. and Komorowski, J. (2016). Selection of significant features using Monte Carlo feature selection. In *Challenges in Computational Statistics and Data Mining*, S. Matwin and J. Mielniczuk, eds. (Springer International Publishing), pp. 25–38.

Delfs, R., Doms, A., Kozlenkov, A., and Schroeder, M. (2004). GoPubMed: Ontology-based literature search applied to Gene Ontology and PubMed. In *Proceedings of German Bioinformatics Conference. LNBI* (Springer), pp. 169–178.

Feng, Z., Jensen, S. M., Messenheimer, D. J., Farhad, M., Neuberger, M., Bifulco, C. B., and Fox, B. A. (2016). Multispectral imaging of T and B cells in murine spleen and tumor. *J. Immunol.*, **196**, 3943.

Fletcher, R. H. and Fletcher, S. W. (2005). *Clinical Epidemiology: The Essentials* (Philadelphia: Lippincott Williams&Wilki).

Hackl, H., Charoentong, P., Finotello, F., and Trajanoski, Z. (2016). Computational genomics tools for dissecting tumour-immune cell interactions. *Nat. Rev. Genet.*, **17**, 441–458.

Kanehisa, M., Furumichi, M., Tanabe, M., Sato, Y., and Morishima, K. (2017). KEGG: New perspectives on genomes, pathways, diseases and drugs. *Nucleic Acids Res.*, **45**, D353–D361.

Krawiec, K. (2002). Genetic programming-based construction of features for machine learning and knowledge discovery tasks. *Genet. Program. Evolvable Mach.*, **3**, 329–343.

Muja, M. and Lowe, D. G. (2014). Scalable nearest neighbor algorithms for high dimensional data. *IEEE Trans. Pattern Anal. Mach. Intell.*, **36**, 2227–2240.

Radivojac, P., Obradovic, Z., Dunker, A. K., and Vucetic, S. (2004). Feature selection filters based on the permutation test. In *Machine Learning: ECML 2004* (Springer, Berlin, Heidelberg), pp. 334–346.

Rand, W. M. (1971). Objective criteria for the evaluation of clustering methods. *J. Am. Stat. Assoc.*, **66**, 846–850.

Schoenmeyer, R., Binnig, G., and Schmidt, G. (2014). Generating image-based diagnostic tests by optimizing image analysis and data mining of co-registered images. US Patent 9159129 B2. Google Patents, http://www.google.com.pg/patents/US9159129.

Schönmeyer, R., Athelogou, M., Schmidt, G., and Binnig, G. (2014). Visualization and navigation platform for co-registered whole tissue slides. In *Bildverarbeitung für die Medizin 2014*, T. M. Deserno, H. Handels, H.-P. Meinzer, and T. Tolxdorff, eds. (Springer Berlin Heidelberg), pp. 13–18.

Schönmeyer, R., Brieu, N., Schaadt, N., Feuerhake, F., Schmidt, G., and Binnig, G. (2015). Automated whole slide analysis of differently stained and co-registered tissue sections. In *Bildverarbeitung Für Die Medizin 2015*, H. Handels, T. M. Deserno, H.-P. Meinzer, and T. Tolxdorff, eds. (Springer Berlin Heidelberg), pp. 407–412.

Stack, E. C., Wang, C., Roman, K. A., and Hoyt, C. C. (2014). Multiplexed immunohistochemistry, imaging, and quantitation: A review, with an assessment of Tyramide signal amplification, multispectral imaging and multiplex analysis. *Adv. Boundaries Mol. Cell. Pathol.*, **70**, 46–58.

Tsodikov, A., Szabo, A., and Jones, D. (2002). Adjustments and measures of differential expression for microarray data. *Bioinformatics*, **18**, 251–260.

Yigitsoy, M., Schmidt, G., and Binnig, G. (2017). Hierarchical patch-based co-registration of differently stained histopathology slides. In *Medical Imaging 2017: Digital Pathology, Proc. SPIE 2017* (Orlando, FL, USA).

Chapter 6

Bioinformatics

Sriram Sridhar,[a] Brandon W. Higgs,[a] and Sonja Althammer[b]
[a]*MedImmune LLC, Gaithersburg, MD, USA*
[b]*Definiens AG, Bernhard-Wicki-Strasse 5, 80636 Munich, Germany*
sridhars@medimmune.com

The advent of omics-based technologies has enhanced the ability to gain specific insights into the molecular pathogenesis of human diseases. The vast amounts of patient-specific data produced by these technologies have been used to augment disease understanding and generate hypotheses around novel diagnostic and prognostic disease biomarkers. Next-generation sequencing (NGS) has brought about the ability to obtain exhaustive individual genetic and transcriptomic profiles across normal and diseased human tissue, even up to the level of single cells within tissue samples. In the field of oncology, the ability to complement genetic and transcriptomic data with clinical characteristics and tissue-based data can facilitate a deeper understanding of clonal heterogeneity within tumors and across patient populations, and enable a more comprehensive exploration of the tumor microenvironment. Transcriptomics allows the assessment of expression changes across a large set of genes, while tissue phenomics elucidates cell-to-cell interactions within biopsy specimens and

---

Tissue Phenomics: Profiling Cancer Patients for Treatment Decisions
Edited by Gerd Binnig, Ralf Huss, and Günter Schmidt
Copyright © 2018 Pan Stanford Publishing Pte. Ltd.
ISBN 978-981-4774-88-8 (Hardcover), 978-1-351-13427-9 (eBook)
www.panstanford.com

measures distributions to quantify specific cell types. Integration of genomic, tissue-based, and clinical information presents promising opportunities to better identify biomarkers of early disease detection, extract molecular and/or cellular patient subsets, and develop a potential prognosis and/or predictive biomarkers of therapeutic treatment response.

6.1 Molecular Technologies: Past to Present

The last quarter century has produced an explosion in the amount of information that can be used to facilitate diagnosis and understanding of the molecular mechanisms driving human disease. Prior to this time, clinicians and scientists had mainly relied on clinical histories and tissue pathology to discern molecular changes that indicated the onset and progression of a disease. Detailed cellular information from patient tissue specimens was typically characterized manually using immunohistochemistry (IHC), while specific genetic and genomic information could be assessed through techniques such as fluorescent *in situ* hybridization (FISH) and polymerase chain reaction (PCR). While these methods provided the ability to examine modulations of selected proteins and individual genes within a snapshot in time, the coordinated actions of multiple genes and proteins within a disease process remained elusive. The onset and completion of the Human Genome Project produced orders of magnitude more information on the genetic (*i.e.*, DNA-based) and transcriptomic (*i.e.*, RNA-based) level in human biospecimens, and new technologies used to evaluate these data in a high-throughput method enabled researchers and clinicians to obtain pan-genomic profiles of normal and diseased tissues from patients in a single experiment. This detailed global view of the genome, coupled with pre-existing clinical and pathological information, allowed scientists to gain a more comprehensive view of the changes that occur across entire biological pathways, including causal mechanisms that may lead to the onset and progression of human disease.

The explosion of biological information on human tissues produced by novel and improved technologies spanned all aspects of the central dogma of biology. The completion of the Human Genome Project produced a complete set of human DNA information to interrogate. Anchoring off of this data and refining structural annotation with various algorithms, scientists could then assemble the human exome, or set of transcribed genes within the genome. As scientists continued to study the genome, they increasingly appreciated the complexity by which this genetic information was organized and regulated. This drove advancements in understanding the mechanisms by which transcription and translation are mediated. Technological improvements enabled researchers to move from single pull-down assays evaluating various transcription factors and cis/trans-acting elements, to large-scale profiling of the epigenetic landscape, including events such as histone deacetylation and methylation. Further, inspection into the non-coding regions of DNA caused biologists to modify pre-existing theories supporting non-functional, or junk DNA comprising a vast majority of the genome; now these non-coding regions are known to contain various transcriptional modifying elements such as binding sites for enhancing and repressing factors, microRNAs (Bartel, 2004), which facilitate post-transcriptional regulation of RNAs, and other families of non-coding RNAs (Cech and Steitz, 2014) with various regulatory functions, which are still being identified (*e.g.*, lincRNAs, snoRNAs). Improved technology has also enabled measurement of biochemical intermediates from a host of cellular mechanisms [*i.e.*, metabolomics (Rochfort, 2005)].

With each new application of *omics* technology, researchers continue to evolve our understanding of the flow and processing of information that drives molecular mechanisms. Improved precision and granularity was also observed in the specimens obtained from patients for analysis. From heterogeneous tissue samples to isolated cell types from tissue or organ systems, to now individual cells (Wu *et al.*, 2014), and circulating cell-free DNA (Wan *et al.*, 2017), the whole genome information described above can be collected at every level of biological structural units. This allows researchers to not only obtain a detailed compartmental view of biological processes but also determine the interplay between systems. By understanding these aspects of normal cellular, tissue, and organ function, we can then begin to understand malfunctions in these systems, which can lead to the onset and progression of a disease.

The abundance of biological information discussed above would not have been made possible without improvements in the technology used to process patient specimens and computational power to process, store, and analyze high-throughput biological data. The ability to systematically analyze large datasets using intricate statistical models requires considerable amounts of computational resources. Prior to the Human Genome Project, molecular mechanisms were restricted to one gene, one protein, one pathway analysis. Due to technical limitations, it was difficult to produce system-level components of signaling pathways or interplay between canonical mechanisms. Applying statistical and mathematical modeling to the increasing amounts of biological data being produced has allowed researchers to model complex and integrated processes in a systematic way. Breakthroughs and improvements in computing have made it possible to apply ever more complex statistical and mathematical approaches to larger amounts of data being produced, and test scientific hypotheses in an exhaustive manner. This has made it possible to incorporate exponentially more information into our understanding of biological processes, while also revealing more possibilities to be further explored experimentally. Taken together, these various improvements in experimental, analytical, and computational technology have enabled researchers to transition from a narrow view of a biological mechanism to a more holistic and integrated view of these systems, which impacts the way we think about how diseases manifest, progress, and can be treated.

6.2 Genomics and Tissue Phenomics

6.2.1 Genomics Data Sources

Sequencing of DNA or RNA from human tissue or blood can generate abundant quantities of molecular information, which can be critical to understanding mechanisms at play in normal physiological processes or in disease. High-throughput sequencing methods, commonly termed NGS, can now be run at substantially lower cost than when initially introduced, and computational approaches to store and analyze these data are becoming increasingly optimized and standardized (Goodwin *et al.*, 2016). Despite these advances, sequencing of whole genomes, exomes, or transcriptomes is still not routinely used in clinical practice (Biesecker *et al.*, 2012; Park *et al.*, 2013; Zhao, 2014). This is, in part, due to the fact that while running whole genome NGS assays has become more streamlined, specimen quality can be a factor and both standardization and speed of interpretation in the analysis are still developing for mass clinical use. For DNA, targeted gene panels with 100 or fewer genes are often sequenced, providing deeper coverage of focused sets of targets at a much reduced cost (Brazma *et al.*, 2001). As a result, depending on the goal of the study, prioritizing targeted sets of genes with focused sequencing assays may adequately address specific questions for which whole genome, exome, or transcriptome sequencing may be extraneous. More comprehensive sequencing efforts can be useful for generating hypotheses on smaller patient cohorts, which can later be tested with targeted panels on larger cohorts. Alternatively, hypothesis generation can be enabled through the use of previously published genomic data.

Analysis of omics data often utilizes information from previously published studies to generate and confirm hypotheses. Much of these data are stored in publicly available resources and include *in vitro* (*i.e.*, data from cell line or cell culture experiments), *in vivo* (*i.e.*, animal model data), and clinical (*i.e.*, data from human samples) datasets. Publicly available databases of omics data are often used to enhance biological understanding of disease mechanisms, conduct exploratory analysis to understand disease mechanisms or characterize patient populations, as well as to validate clinical observations through the use of additional data. The work of the MIAME consortium (Brazma *et al.*, 2001) set up guidelines for the mandatory submission of microarray data for gene expression studies, which were published and thus began the process of standardizing expression data and metadata associated with transcriptomic studies. The MIAME criteria are utilized in the two main public transcriptomic databases, the Gene Expression Omnibus (GEO) (Barrett *et al.*, 2013) and ArrayExpress (Kolesnikov *et al.*, 2015), allowing investigators access to raw and normalized gene expression data with a minimum amount of metadata to replicate analysis of transcriptomic studies. With the advent of high-throughput sequencing approaches to interrogate the transcriptome, additional data repositories such as the Sequence Read Archive (Bethesda, 2009) were created to house the larger sets of raw data associated with these platforms. In addition to transcriptomic databases, resources housing published genomic data have also been created along the same lines to facilitate re-analysis and discovery of novel genetic variants across published studies. dbSNP and dbGAP are prominent examples of databases that house DNA sequencing and SNP-based studies. These repositories enable researchers to amass large amounts of data related to specific areas of study or toward addressing specific clinical questions, without having to generate new data to answer specific questions. The information provided by these resources often provides an initial look into addressing questions about disease biology as part of exploratory or confirmatory analyses.

A major advantage of resources such as GEO and ArrayExpress is the allowance of investigators to explore biological questions across diverse sets of information without having to perform additional experiments. Utilizing previously published information to answer questions, especially when the information is as rich as transcriptomic data, can be a significant cost savings for researchers while also being a valuable resource for generating new hypotheses and validating existing ones. While the data in these resources are abundant and often accompanied by various useful pieces of metadata, care must be taken in how information from these sources is handled. Because of differences in technical platforms used (*i.e.*, types of microarrays used), cell or tissue types assayed, or metadata available to cross-compare, it may be appropriate to look at datasets individually as opposed to combining them across studies. While being able to combine multiple datasets increases sample sizes and statistical power, analyzing data across disparate studies can introduce technical biases, which may confound trends observed in the data.

In addition to the genomic and transcriptomic repositories mentioned above, resources have also been created to facilitate discovery focused on specific disease areas. The Cancer Genome Atlas (TCGA) (Tomczak *et al.*, 2015) is a prominent example of this type of resource. TCGA houses proteomic, genomic, transcriptomic, and clinical metadata across a variety of cancer indications, mostly from patient tumors not exposed to treatment. A major goal of the resource is to serve as a hub for mining cancer genomic and transcriptomic information. By storing this information on patients, researchers using TCGA data can carry out analyses such as class discovery or identify patterns of interest among genes or pathways within specific cancers or patient sub-populations within or across indications. With genomic information serving as an initial driver, transcriptomic data providing a cross-sectional view of molecular activity, and clinical information indicating disease phenotypes, researchers can begin to investigate the causal mechanism driving the onset and progression of different malignancies.

Integration of transcriptomic datasets can also be used to estimate the abundance of cell-specific populations or activity of specific biological pathways within biosamples. Methods such as gene set variation analysis (GSVA) (Hänzelmann *et al.*, 2013) and CIBERSORT (Newman *et al.*, 2015) utilize previously published data consisting of signatures, or sets of genes, which are representative of cell populations (*e.g.*, immune cell populations) or signaling pathways. Various databases such as MSigDB or BioCarta compile signatures for use in transcriptomic analysis. Individual studies also make use of published datasets (Grouse *et al.*, 2001; Subramanian *et al.*, 2005) to extract signatures that are representative of cell types (Allantaz *et al.*, 2012; Charoentong *et al.*, 2017) and can be used to estimate abundance or proportions of cells within biosamples. The algorithms mentioned utilize gene signatures by generating a summary of the expression of groups of genes into a composite score, which can be calculated per sample or per patient. The end result is a set of scores for cell types or pathways, per patient, which can then be compared across larger cohorts. These methods, from gene set compilation to signature scoring, all utilize transcriptomic data, either previously published or prospective, to enhance the granularity of information one can derive on individual patients or samples.

All the examples mentioned above highlight how diligently curated public data repositories can be harnessed to enhance the understanding of disease biology and molecular mechanisms prevalent in individual studies. Researchers can now take advantage of the variety and depth of previously published research in a systematic fashion to generate novel hypotheses and validate existing hypotheses. The increased transparency behind the datasets published in these repositories, including the available metadata published along with them, allows researchers to better discern intricacies in specific diseases or patient groups, which can help them to better coordinate prospective studies. With the advent of streaming data sources providing more granular information on individual physiology, these resources will continue to evolve to integrate more diverse information on individuals, moving toward a more targeted and personalized understanding of disease biology. Initiatives such as NIHs Precision Medicine Initiative (Collins and Varmus, 2015) represent ambitious efforts to manage and integrate streaming data across large sets of individuals, with the goal of utilizing this diverse information in a meaningful way to better understand human physiology and treat human disease.

6.2.2 The Art of Image Mining

Tissue phenes, which are meaningful features extracted from histological slides, have been successfully associated to the outcome of cancer patients. Before providing examples, we first describe the automated extraction of phenes from IHC-stained and digitized tumor tissue sections—one of the most complex and information-rich data types a scientist may encounter. While the content of tissue slides does not directly reveal information related to the genotype, it captures information about the phenotype, described by the actual quantification of numerous cell populations, the interplay between them, and the geographical landscape of a snapshot of the tumor microenvironment.

The quantification of marker-positive cell populations, which has traditionally been conducted with manual assessment, has become automated by sophisticated algorithms with the capability of achieving highly accurate and reproducible results (Chapter 4). One primary motivation for this advancement in the technology is that manually counting cells, quantifying areas, capturing morphological characteristics of cells, and measuring exact cell-to-cell distances do not provide a practical option in clinical application and, pragmatically, simply not possible. Not all aforementioned features might be extractable from a given dataset. The art of image mining combines the both relevant and feasible analyses with appropriate algorithms and complex image processing/analysis with clinical outcomes. The feasibility of the feature extraction highly depends on the quality of the tissue and the objects resulting from the image analysis, as well as of the proximity of the markers under evaluation in a three-dimensional space, or more specifically, the distances of the section, in the event that co-registration was used.

6.2.2.1 Cell-to-cell distances

Investigating the spatial distribution of one or more cell populations may be motivated by the hypothesis that the probability of cell interactions is linked to cell-to-cell distances. If a conclusion about direct cell contact is the aim, we might want to consider a distance in the range of 20 µm and, therefore, in an ideal setting, one tissue section contains all markers of interest (multiplex). There are different methods to calculate proximity, and some of them are demonstrated in Fig. 6.1. One example is a proportional proximity score, which can be calculated by the ratio of close objects versus all objects of a certain type. This approach can be applied considering one or multiple cell populations as well as other objects such as cell clusters. Parameters including the definition of the objects and the lengths of the radii can be adapted to the hypothesis.

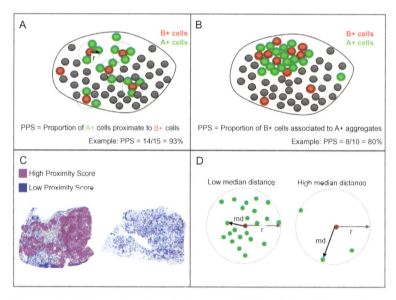

Figure 6.1 (A) Proportional proximity score (PPS) is calculated by identifying A+ cells that are within a radius r around the object of interest (B+ cells) and dividing their number by the total number of A+ cells. (B) Similarly, a PPS for one cell population to a cluster of another cell population can be calculated. (C) Heatmap examples displaying high and low proximity scores. (D) Proximity scores can be derived by averaging over the number of cells in a circle with radius r or over the median distances (md) to those circumferential cells.

When co-registered sections are used, the risk of falsely assuming proximity might become high depending on the tolerated distance, the proximity of the sections, as well as on the expected error of the co-registration algorithm. If direct cell-to-cell interactions are not the focus of investigation, but a rough idea about the proximity of cell objects is desired, a simple statistic of the median distances of those objects might be an appropriate choice. To gain confidence in the calculations, results can be compared to what is expected by chance through shuffling the object labels.

6.2.2.2 Quantifying cell populations in histological regions

The role of stroma and the cells within this region within the tumor microenvironment is an active area of research. In order to test different hypotheses, the distinction between stroma and epithelium is critical. Sophisticated image analysis methods can distinguish between these histological regions and enable a quantification of the embedded cell populations, for example, tumor-infiltrating lymphocytes (TILs) in stroma versus epithelium. The quality of the detection of these regions may depend on the quality of the assay, which is an important factor to consider before interpreting the results.

Regions of interest can also be annotated manually and captured by the tissue phenomics platform. Examples are the tumor center (TC) and the invasive margin (IM), which are established regions of interest. Figure 6.2 shows how an image analysis algorithm assists the pathologist: For the IM designation, a line simply needs to be drawn and then grown in either direction to create an IM region. While allowing for regional quantifications, interesting patterns can be identified, such as a T-cell *jam* at the IM as opposed to a heavily infiltrated tumor (Fig. 6.3A). Heatmaps can help to visualize those patterns as they can be employed to amplify the signal (Fig. 6.3B). Heatmap mining is a substantial part of image mining as is elaborately explained in Section 5.2.2.

It is important to mention that image mining works best when it is closely connected to expertise from pathology, image analysis, and quality and control processes. This is described as the tissue phenomics loop in Section 5.5.

Genomics and Tissue Phenomics | **141**

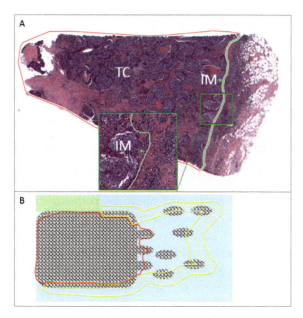

Figure 6.2 Tumor center (TC) and IM annotations: (A) Annotations of TC (red) and IM (green) on an H&E-stained tissue section. (B) Diagram shows how the IM region is generated by growing the annotated IM line by a defined distance (*e.g.*, 250 µm) in both directions.

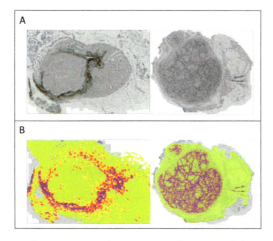

Figure 6.3 T cells occurring in distinct regional patterns: (A) Jam of T cells at the IM (left) versus a TC infiltrated by T cells (right). (B) Corresponding heatmaps for visualization and analysis.

6.2.2.3 Tissue phene and survival

Numerous researchers have identified associations between tissue phenes and patient survival. Gooden *et al.* applied a meta-analysis to review a number of studies that reported a prognostic influence of various subsets of TILs in cancer and confirmed the positive effect of CD3(+) and CD8(+) TILs (Gooden *et al.*, 2011). Further, a famous study was conducted on these subsets of TILs in separate histological regions: Galon and colleagues have introduced a cancer classification by a so-called Immunoscore (Galon *et al.*, 2012). High densities of CD8(+) and CD3(+) lymphocytes within the TC and the IM have been associated with longer survival of patients with colorectal cancer. The discovery of Galon *et al.* resulted in a product called Immunoscore®, which is intended to support clinicians in assessing the prognosis of primary colon cancer patients. The Immunoscore has demonstrated a greater prognostic value than tumor staging, and the ability to retrieve the score automatically from digitized images represents a turning point in the era of tissue phenomics. Another report investigated TIL subsets in more detail in gastric cancer and provided evidence for the existence of an ideal distance between CD8(+) and FOXP3(+) T cells to effectively provoke an immune response (Feichtenbeiner *et al.*, 2014). This mechanism is supported by the ability of regulatory T cells to reduce effector T-cell function by cell-to-cell interaction and cytokine secretion (Piccirillo and Thornton, 2004). A final example, the spatial relationship between T cells and tumor cells, represents another very active area of investigation in the field of tissue phenomics. While studying the spatial distribution of T cells and cancer cells in pancreatic cancer, Carstens *et al.* discovered a relationship between the proximity of TILs to cancer cells and a favorable outcome of the patients (Carstens *et al.*, 2017).

6.2.3 Power of Integrative Approaches

The methods described in the previous sections present aspects of comprehensive molecular or cellular characterizations of disease or normal tissue from patients. In the setting of cancer, the ability to assess genomic, transcriptomic, and protein-based markers in tumor samples from patients provides a comprehensive profile of a patient's tumor microenvironment. In the past, these pieces of information may have been used in isolation to provide incremental information on individuals, which may be acted on independently. The ability to integrate large sets of information in a meaningful way can help to understand an individual's specific disease course prospectively and in real time and help guide diagnosis, treatment, and prognosis decisions.

For example, large-scale transcriptome screening can be leveraged to prioritize IHC protein markers while designing a study. Even though high quantities of mRNA do not necessarily lead to high quantities of proteins (due to post-transcriptional regulation, *etc.*), the expression level of a gene can provide supporting evidence to include the corresponding protein as a target in the IHC protocol. Similarly, integrative approaches can be very helpful to support findings while employing different technologies. IHC image analysis results can be supported by gene expression data from the same set of samples or even from an independent set based on the assumption that the considered datasets consist of representative groups of subjects. Further, cellular markers measured by IHC can be used to deconvolute mRNA expression profiles. Since gene expression is typically measured from heterogeneous tissue matrices, immunostaining for certain cell-specific markers can inform cell-specific mRNA expression patterns. Clinical information, including patient survival, disease stage, and other blood or cellular markers can be incorporated in the interpretation of genomic, transcriptomic, or IHC imaging data to understand which patients with elevated or decreased levels of genomic or tissue phenomic markers have poorer prognosis, or which clinical factors are enriched among patients with specific molecular profiles. This level of integration can affect treatment decisions beyond traditional approaches that rely solely on clinical factors (such as disease severity or tumor stage) since it does not necessarily rely on previously defined categorizations of disease. While cancer treatment is moving toward personalized medicine, integrative approaches will be indispensable as they facilitate the generation of combined biomarkers to refine definitions of cancer subclasses.

This capability of integrating genomic data with digital IHC imaging to inform clinical decisions hinges on biologically driven hypotheses. While these technologies assess seemingly disparate aspects of cellular or molecular processes, they exhibit unique strengths that complement each other and result, if combined, into a comprehensive basis for downstream analysis. The methods used to analyze and interpret the data can be quite similar, which can aid in effectively assimilating imaging and genomic data to better understand disease processes and ultimately benefit patients.

6.3 Analytical Approaches for Image and Genomics Data

6.3.1 Data Handling Requires Similar Methods

Digital imaging methods and genomics approaches both produce large amounts of data, which usually require computational algorithms to process, mine, and interpret. Many of the methods commonly used to analyze digital pathology data are also used to assess genomic data. Examples of some of these commonly used approaches include, but are not limited to, signal/image processing, machine learning, classification/clustering, pattern recognition, and integration with clinical information. Through the use of digital pathology methods outlined in the previous chapters, histological images can essentially be codified into numerical data, which can be analyzed in a similar manner as genomic information. This codified data can be assessed using statistical methods to comprehensively characterize individual and group-wide trends in data, which can be then used to identify discerning characteristics about individual or groups of samples. The common thread between these media is the representation of biological information (tissue/cell-level and gene-level) as numerical data, allowing common approaches to be applied to both. Genomic (DNA-based information), transcriptomic (RNA), or other omic data (*e.g.*, metabalomic, proteomic) typically consist of sets of features (mRNA molecules, variant alleles, protein expression, *etc.*) that can be combined with features derived from image analysis, represented for individual samples across larger populations. Patient data are typically complemented with additional meta-information (*e.g.*, age, gender, ethnicity, disease status), which can be used to create groups across which omic-based features can be compared. Discerning trends can be measured with statistical confidence when comparing groups to one another, or even when observing individuals within a group (creating subgroups). Approaches such as machine learning provide methods for identifying which features are associated with specific clinical or biological outcomes. In addition, certain characteristics can predict groups of patients who may fit into a specific disease category, have better/worse prognosis, or respond better to a specific therapy.

While much can be gleaned from the analysis and integration of these high-throughput datasets, there are also challenges that need to be considered in their interpretation. Datasets derived from digitized image analysis and omics approaches are typically high dimensional (many features) with relatively few observations (patients). As a result, the statistical models used to discern trends in data need to be adequately trained, accounting for false positive discoveries through methods such as cross-validation and permutation tests. This step is essential for confirming findings across datasets and applying methods to prospective studies. Failure to do this correctly can result in overfitting, where a statistical model used to describe data is minimally generalizable and typically applicable only to the original study analyzed, but not to other studies that may have subtle differences that cannot be explained with the current approach. Being able to account for these factors is essential to fully utilizing the information provided by genomic and digital IHC approaches.

6.3.2 Enhancing Confidence in a Discovery

When applied to tissue phenomics and/or other omics data, machine learning and data mining approaches provide powerful tools to gain insights into biological data and to convert them into knowledge about the disease and its underlying mechanisms. Once a pattern is elucidated, it is important to validate this across similar datasets to increase confidence in the likelihood that the finding is not due to random chance. Iterating back to original data is a common approach that can help to provide confidence in the discovery. For example, permutation tests evaluate the robustness of features selected to explain trends in a correlative pattern. This approach ensures that the features selected perform better than randomly selected sets in identifying trends of interest. Cross-validation approaches also help determine the robustness of statistical models used to explain data.

The underlying assumption is that training data are representative of the entire population. While this is often an underlying assumption, cross-validation represents a robust way to exploit the available data and eliminate features that are unstable using training data. The general approach underlying cross-validation involves splitting the data into a number of (N) folds. The $N-1$ folds are used for training the model parameters and 1 fold is used for testing. This procedure is conducted $N-1$ times such that all instances of all folds are classified once. The performance of the model is calculated by comparing the prediction with the actual class. Then the procedure is repeated over multiple iterations until the variation in the model performance stabilizes. The observed variation of the model performances gives an estimate of how generalizable the models are and can be used for prioritization. Cross-validation and other tests of robustness help to provide confidence in discovered features. If the predicted outcomes remain consistent across multiple subsets or variations of the original dataset, more confidence can be placed in the features and model used to explain or predict specific outcomes.

Ideally, if one seeks to gain confidence in a given model, it is best to test the model on independent datasets. Here the same approach is used to gain initial confidence in a model (*e.g.*, permutation tests, cross-validation) and can be applied to prospective or independent datasets. With this approach, the goal is to determine if the model used to explain trends from an initial set of data can be applied to other similar but independent datasets. If so, it would increase confidence that the features used to explain specific trends can be reproducibly detected in larger cohorts and are, therefore, adequately robust.

From data handling to analysis, the workflow described above is relevant to both genomic and imaging data, underscoring the similarities in how these sets of information are handled. Each approach provides sets of features that can be used to identify or explain trends in patient populations based on the mechanisms associated with their conditions. Constant refinement of the statistical models used to explain integrated imaging and genomic data from the features inputted to the interpretation of results can help to identify characteristics that best explain trends associated with patient outcomes. The ability to apply these approaches at scale, where larger sets of data (where available) can be examined in a similar fashion, can increase the confidence in the components of tissue phenomics or genomic data, which best explain outcomes of interest. Integration of these sources of data using approaches that can be easily implemented across centers and interpreted in meaningful ways will help us to better understand how to tailor treatments to individuals based on robust information, which has been rigorously tested.

6.4 Examples of Genomics or IHC Biomarkers in Clinical Practice

6.4.1 Biomarker Background

The NIH working definitions of a biomarker is "a characteristic that is objectively measured and evaluated as an indicator of normal biological processes, pathogenic processes, or pharmacologic responses to a therapeutic intervention." Examples of such biomarkers include the mutational status of a gene (*e.g.*, KRAS), protein levels (*e.g.*, HER2), gene/gene signature expression levels (PAM50), or metabolite levels (*e.g.*, mFBP), as well as combinations of factors. One major utility of biomarkers is the ability to help tailor a patient's diagnosis, prognosis, or treatment approach based on individualized measures. When extended to larger populations, biomarkers can be used to refine disease diagnosis, predict patient outcomes, or inform potential response to therapy. The utilization of biomarkers in clinical trial design has become common practice as it can aid to increase the probability of the trial's success. Biomarkers can be categorized as prognostic or predictive, depending on whether they are associated with a patient's outcome regardless of treatment or on a specific treatment setting, respectively. In both cases, the underlying question is to identify a subpopulation that is associated with a particularly good outcome, for example, better therapy response, longer survival or absence of severe adverse events. In the context of companion or complementary diagnostics, biomarkers play a crucial role, as they may be used to determine if a specific treatment should be given or not. Therefore, it is important that the biomarker is shown to be predictive for the success of the specific treatment. Here, the distinction between prognostic and predictive biomarkers becomes crucial. If biomarker-positive patients live longer regardless of the treatment, the biomarker likely has a stronger prognostic effect than predictive effect, and the rationale for the development of a companion or complementary diagnostic may need to be re-evaluated.

When it comes to the discovery of biomarkers, one can distinguish between hypothesis-driven discovery, which is guided by scientific know-how, and discovery-based discovery, which is guided by the data. While each approach has its own advantages, combined approaches can be very useful (McDermott *et al.*, 2013). As the technologies that assess potential biomarker candidates have matured (*e.g.*, gene expression levels, IHC), the information available to generate biomarkers has evolved over time. The following examples highlight specific genomic and IHC-based biomarkers that have come to prominence in recent years. The novelty and utility of these biomarkers open numerous possibilities of integrating this information to increase their individual impact.

6.4.2 Genomics and IHC to Guide Prognosis or Diagnosis

Diagnosis and prognosis of breast cancer has benefited greatly from advancements in both IHC and genomic technology. Enhanced understanding of the molecular markers underlying tumorigenesis of breast tissue has led to the discovery of the importance of surface receptor expression in determining the aggressiveness of certain breast cancers. Specifically, IHC analysis has shown that the estrogen and progesterone hormone receptors [ER (Harvey *et al.*, 1999) and PR, respectively] are prominent among breast cancers with poor prognosis. Among the potential causes for aberrant receptor activity are mutations in genes coding for these hormone receptors. Mutations in the HER2 growth factor receptor are associated with excessive cell growth *in vitro*. When assessing the potential link between this cell growth and cancer, it was reported that excessive levels of HER2 were present in large numbers of breast cancer samples and that these HER2-positive cancers were typically more aggressive growing tumors (Slamon *et al.*, 1987). With ER and PR, HER2 presented a third tumor surface receptor whose status was indicative of more aggressively growing and potentially treatment-resistant cancers. As surface receptors, cancers with high levels of ER, PR, or HER2 can be amenable to direct blockade by antibodies *(e.g.,* Herceptin for HER2-positive cancers) or indirect blockade by small molecules *(e.g.,* tamoxifen for ER-positive cancers), resulting in reduction in tumor growth. Since levels of these receptors can be measured by IHC from breast tissue biopsies, the status of these receptors serves as prognostic markers for determining amenability of certain tumors with high hormone or HER2 levels to receptor blocking drugs. In addition, statuses of these receptors can also be used to molecularly classify breast cancer patients, where patients who are positive for one or more surface receptors (ER, PR, and/ or HER2) belong to one class of cancers, while patients who are negative for all three receptors *(i.e.,* triple negative) belong to another. These classifications, in addition to beginning to elucidate potential biological mechanisms for disease pathogenesis, have helped to expedite treatment options for patients and ultimately improve overall and progression-free survival.

The use of IHC to identify molecular subclasses of breast cancers, which could then be targeted for specialized treatment, raised the possibility of assessing molecular activity in breast tissue, which may manifest prior to being visible histologically, but still be associated with diagnosis or progression of disease. Transcriptomic analysis of histologically normal tissue has been shown in many instances to be altered in individuals who eventually go on to develop diseases, including cancers. Based on this rationale, it is possible to detect molecular changes in the pre-malignant or early-stage disease, which may be indicative of the molecular activity seen in the later-stage disease. The MammaPrint© test for breast cancer is an example of this concept (Drukker *et al.,* 2013). The initial work that led to the development of this test had shown that based on microarray-based gene expression analysis of breast tumor tissue, there was a panel of genes whose expression was shown to be associated with metastasizing tumors (van de Vijver *et al.,* 2002). Based on this finding, probabilistic models were generated to determine the likelihood that given a certain pattern of expression of a set of genes, a patient's tumor would metastasize. Using these models, patient tumors could then be assessed prospectively to determine if the panel of genes could accurately determine the likelihood of a tumor metastasizing. The 70-gene MammaPrint panel, which was originally identified by transcriptomic analysis of formalin-fixed paraffin-embedded (FFPE) tumor tissue samples, is currently an US

Food and Drug Administration (FDA)-approved (as of 2007) method for helping to determine the risk of metastases and subsequently to guide treatment options for patients with potentially aggressive breast cancers.

The use of IHC to determine receptor status in breast cancer and the MammaPrint panel for determining risk of metastases are examples of histology and genomic information being used to better understand cancer genesis and progression, and ultimately to guide treatment of patients with life-threatening cancers. Digital analysis of pathology data makes it possible to more systematically assess tumor histology and molecular characteristics of tumors. Improved data management and integration make it possible to meaningfully link genomic data with clinical information, including imaging data, to gain a more comprehensive and potentially exhaustive understanding of types of cancers, drivers of their progression, and options for expediting their treatment. IHC of cell receptor status and transcriptomics are individual factors that present aspects of disease that can be accounted for to understand pathogenesis and are currently used to guide treatment. The ability to now assess more information within an individual and across larger cohorts of patients will allow us to better link these pieces of information into one coherent picture of disease onset and progression.

With this more holistic understanding, it will be clearer how best to modulate individual factors to create customized and more sustainable modes of treatment that not only target individuals, but can be used to characterize large groups of individuals, making disease diagnosis and treatment more streamlined.

6.4.3 Patient Stratification: Genomics and IHC to Identify Patient Subsets for Treatment

In addition to informing disease prognosis, genomic and IHC-based biomarkers can also be used to determine subsets of patients who may respond to specific therapies. Many of these biomarkers have been extensively categorized and reviewed (Herbst *et al.*, 2016; Ludwig and Weinstein, 2005; Mehta *et al.*, 2010). ER and PR status in breast cancer is used to predict sensitivity of patients to hormone therapy, while HER2 receptor status predicts responsiveness of patients to Herceptin treatment. Point mutations or translocations in BCR-ABL infer sensitivity of CML patients to Imatinib treatment. Melanoma patients with mutations in BRAF are more likely to respond to Zelboraf treatment (Hyman *et al.*, 2015). In the case of ER/PR and HER2 status, these biomarkers are both predictive of response to therapy as well as indicative of poor prognosis, illustrating how biomarkers can be used for different purposes to facilitate disease understanding and inform clinical treatment. For determining status of protein markers or receptors, IHC has been commonly used; however, biomarkers that have been more recently developed have begun to incorporate genomic data from sources including FISH, PCR, as well as gene expression and sequencing information (DNA- and RNA-based) to measure biomarkers and select patients earlier in their disease progression and using increasingly minimally invasive sampling techniques.

With the increased utility of genomic information, biomarkers now can incorporate several readouts (*i.e.*, multiple genes) to predict response to therapies. Recent immuno-oncology trials have utilized IHC of checkpoint inhibitors to inform patient selection for various clinical trials. PD-L1 status, as measured by IHC, has been shown to be associated to response to PD-L1 blockage for treatment of non-small cell lung cancer (NSCLC) (Herbst *et al.*, 2016). In addition to PD-L1 IHC, which does predict a percentage of patients likely to respond to immune checkpoint blockade, current transcriptomic biomarkers are also being developed, which, in combination with IHC tests, can further stratify patients and affect selection for clinical trials. Genes such as interferon gamma (IFNG) in combination with PD-L1 protein as measured by IHC has been shown to identify NSCLC patients with improved clinical benefit from durvalumab, an anti-PD-L1 molecule developed by AstraZeneca/MedImmune (Higgs *et al.*, 2016). In addition, a randomized study evaluating atezolizumab in NSCLC patients (POPLAR), another anti-PD-L1 therapy, showed that mRNAs associated with effector T-cell function could select a subset of patients with improved survival (Williams *et al.*, 2016). This biomarker was evaluated with PD-L1 protein for predictive purposes as well. In fact, utility of IHC data in predicting sensitivity to therapies can impact the success of clinical trials. Two recent sets of trials (KEYNOTE-024 by Merck and CheckMate-026 by BMS) for the treatment of NSCLC with two different PD-1 blockers (pembrolizumab by Merck and nivolumab by BMS) were recently conducted, both of which used PD-L1 IHC as a biomarker for selecting patients for treatment. However, only one trial (KEYNOTE-024) showed significant response rates among PD-L1-high subjects, while the other did not. Closer investigation into the trials suggests that the criteria around magnitude of PD-L1 levels used in the biomarker evaluation and subsequent patient selection may have played a role in the success of pembrolizumab versus nivolumab in meeting its primary trial endpoints (Reck *et al.*, 2016).

Moving forward, the utility of genomic information and integration with IHC data will continue to play a role in the development of predictive and prognostic biomarkers for therapy. There will continue to be a push in the field to develop these biomarkers using as minimally invasive patient sampling as possible, and with the goal of detection of susceptibility and sensitivity to therapy as early in disease progression as possible.

Acknowledgments

In addition to the coauthors of the book, we would like to thank Tze Heng Tan, Andreas Spitzmüller, Moritz Widmaier, Aleksandra Zuraw, Song Wu, and Koustubh Ranade for contributing ideas and illustrations to this chapter. Special thanks go to the development team of the tissue phenomics platform, particularly to Arno Schäpe and Victor Matvienko.

References

Allantaz, F., Cheng, D. T., Bergauer, T., Ravindran, P., Rossier, M. F., Ebeling, M., Badi, L., Reis, B., Bitter, H., D'Asaro, M., *et al.* (2012). Expression profiling of human immune cell subsets identifies miRNA-mRNA regulatory relationships correlated with cell type specific expression. *PloS One*, **7**, e29979.

Barrett, T., Wilhite, S. E., Ledoux, P., Evangelista, C., Kim, I. F., Tomashevsky, M., Marshall, K. A., Phillippy, K. H., Sherman, P. M., Holko, M., *et al.* (2013). NCBI GEO: Archive for functional genomics data sets—update. *Nucleic Acids Res.*, **41**, D991–D995.

Bartel, D. P. (2004). MicroRNAs: Genomics, biogenesis, mechanism, and function. *Cell*, **116**, 281–297.

Bethesda. (2009). GaP FAQ Archive (National Center for Biotechnology Information, US).

Biesecker, L. G., Burke, W., Kohane, I., Plon, S. E., and Zimmern, R. (2012). Next generation sequencing in the clinic: Are we ready? *Nat. Rev. Genet.*, **13**, 818–824.

Brazma, A., Hingamp, P., Quackenbush, J., Sherlock, G., Spellman, P., Stoeckert, C., Aach, J., Ansorge, W., Ball, C. A., Causton, H. C., *et al.* (2001). Minimum information about a microarray experiment (MIAME): Toward standards for microarray data. *Nat. Genet.*, **29**, 365–371.

Carstens, J. L., Correa de Sampaio, P., Yang, D., Barua, S., Wang, H., Rao, A., Allison, J. P., LeBleu, V. S., and Kalluri, R. (2017). Spatial computation of intratumoral T cells correlates with survival of patients with pancreatic cancer. *Nat. Commun.*, **8**, 15095.

Cech, T. R. and Steitz, J. A. (2014). The noncoding RNA revolution-trashing old rules to forge new ones. *Cell*, **157**, 77–94.

Charoentong, P., Finotello, F., Angelova, M., Mayer, C., Efremova, M., Rieder, D., Hackl, H., and Trajanoski, Z. (2017). Pan-cancer immunogenomic analyses reveal genotype-immunophenotype relationships and predictors of response to checkpoint blockade. *Cell Rep.*, **18**, 248–262.

Collins, F. S. and Varmus, H. (2015). A new initiative on precision medicine. *N. Engl. J. Med.*, **372**, 793–795.

Drukker, C. A., Bueno-de-Mesquita, J. M., Retèl, V. P., van Harten, W. H., van Tinteren, H., Wesseling, J., Roumen, R. M. H., Knauer, M., van 't Veer, L. J., Sonke, G. S., *et al.* (2013). A prospective evaluation of a breast cancer prognosis signature in the observational RASTER study. *Int. J. Cancer*, **133**, 929–936.

Feichtenbeiner, A., Haas, M., Büttner, M., Grabenbauer, G. G., Fietkau, R., and Distel, L. V. (2014). Critical role of spatial interaction between CD8[+] and Foxp3[+] cells in human gastric cancer: The distance matters. *Cancer Immunol. Immunother. CII*, **63**, 111–119.

Galon, J., Pagès, F., Marincola, F. M., Angell, H. K., Thurin, M., Lugli, A., Zlobec, I., Berger, A., Bifulco, C., Botti, G., *et al.* (2012). Cancer classification using the Immunoscore: A worldwide task force. *J. Transl. Med.*, **10**, 205.

Gooden, M. J. M., de Bock, G. H., Leffers, N., Daemen, T., and Nijman, H. W. (2011). The prognostic influence of tumour-infiltrating lymphocytes in cancer: A systematic review with meta-analysis. *Br. J. Cancer*, **105**, 93–103.

Goodwin, S., McPherson, J. D., and McCombie, W. R. (2016). Coming of age: Ten years of next-generation sequencing technologies. *Nat. Rev. Genet.*, **17**, 333–351.

Grouse, L. H., Munson, P. J., and Nelson, P. S. (2001). Sequence databases and microarrays as tools for identifying prostate cancer biomarkers. *Urology*, **57**, 154–159.

Hänzelmann, S., Castelo, R., and Guinney, J. (2013). GSVA: Gene set variation analysis for microarray and RNA-seq data. *BMC Bioinformatics*, **14**, 7.

Harvey, J. M., Clark, G. M., Osborne, C. K., and Allred, D. C. (1999). Estrogen receptor status by immunohistochemistry is superior to the ligand-binding assay for predicting response to adjuvant endocrine therapy in breast cancer. *J. Clin. Oncol. Off. J. Am. Soc. Clin. Oncol.*, **17**, 1474–1481.

Herbst, R. S., Baas, P., Kim, D.-W., Felip, E., Pérez-Gracia, J. L., Han, J.-Y., Molina, J., Kim, J.-H., Arvis, C. D., Ahn, M.-J., *et al.* (2016). Pembrolizumab versus docetaxel for previously treated, PD-L1-positive, advanced non-small-cell lung cancer (KEYNOTE-010): A randomised controlled trial. *The Lancet*, **387**, 1540–1550.

Higgs, B. W., Morehouse, C., Streicher, K., Rebelatto, M. C., Steele, K., Jin, X., Pilataxi, F., Brohawn, P. Z., Blake-Haskins, J. A., Gupta, A. K., *et al.* (2016). Relationship of baseline tumoral IFNγ mRNA and PD-L1 protein expression to overall survival in durvalumab-treated NSCLC patients. *J. Clin. Oncol.*, **34**, 3036–3036.

Hyman, D. M., Puzanov, I., Subbiah, V., Faris, J. E., Chau, I., Blay, J.-Y., Wolf, J., Raje, N. S., Diamond, E. L., Hollebecque, A., *et al.* (2015). Vemurafenib in multiple nonmelanoma cancers with BRAF V600 mutations. *N. Engl. J. Med.*, **373**, 726–736.

Kolesnikov, N., Hastings, E., Keays, M., Melnichuk, O., Tang, Y. A., Williams. E., Dylag, M., Kurbatova, N., Brandizi, M., Burdett, T., *et al.* (2015). ArrayExpress update: Simplifying data submissions. *Nucleic Acids Res.*, **43**, D1113–6.

Ludwig, J. A. and Weinstein, J. N. (2005). Biomarkers in cancer staging, prognosis and treatment selection. *Nat. Rev. Cancer*, **5**, 845–856.

McDermott, J. E., Wang, J., Mitchell, H., Webb-Robertson, B.-J., Hafen, R., Ramey, J., and Rodland, K. D. (2013). Challenges in biomarker discovery: Combining expert insights with statistical analysis of complex omics data. *Expert Opin. Med. Diagn.*, **7**, 37–51.

Mehta, S., Shelling, A., Muthukaruppan, A., Lasham, A., Blenkiron, C., Laking, G., and Print, C. (2010). Predictive and prognostic molecular markers for cancer medicine. *Ther. Adv. Med. Oncol.*, **2**, 125–148.

Newman, A. M., Liu, C. L., Green, M. R., Gentles, A. J., Feng, W., Xu, Y., Hoang, C. D., Diehn, M., and Alizadeh, A. A. (2015). Robust enumeration of cell subsets from tissue expression profiles. *Nat. Methods*, **12**, 453–457.

Park, J. Y., Kricka, L. J., and Fortina, P. (2013). Next-generation sequencing in the clinic. *Nat. Biotechnol.*, **31**, 990–992.

Piccirillo, C. A. and Thornton, A. M. (2004). Cornerstone of peripheral tolerance: Naturally occurring CD4+CD25+ regulatory T cells. *Trends Immunol.*, **25**, 374–380.

Reck, M., Rodríguez-Abreu, D., Robinson, A. G., Hui, R., Csőszi, T., Fülöp, A., Gottfried, M., Peled, N., Tafreshi, A., Cuffe, S., *et al.* (2016). Pembrolizumab versus chemotherapy for PD-L1–positive non–small-cell lung cancer. *N. Engl. J. Med.*, **375**, 1823–1833.

Rochfort, S. (2005). Metabolomics reviewed: A new "omics" platform technology for systems biology and implications for natural products research. *J. Nat. Prod.*, **68**, 1813–1820.

Slamon, D. J., Clark, G. M., Wong, S. G., Levin, W. J., Ullrich, A., and McGuire, W. L. (1987). Human breast cancer: Correlation of relapse and survival with amplification of the HER-2/neu oncogene. *Science*, **235**, 177–182.

Subramanian, A., Tamayo, P., Mootha, V. K., Mukherjee, S., Ebert, B. L., Gillette, M. A., Paulovich, A., Pomeroy, S. L., Golub, T. R., Lander, E. S., *et al.* (2005). Gene set enrichment analysis: A knowledge-based approach for interpreting genome-wide expression profiles. *Proc. Natl. Acad. Sci.*, **102**, 15545–15550.

Tomczak, K., Czerwińska, P., and Wiznerowicz, M. (2015). The Cancer Genome Atlas (TCGA): An immeasurable source of knowledge. *Contemp. Oncol.*, **19**, A68–A77.

van de Vijver, M. J., He, Y. D., van 't Veer, L. J., Dai, H., Hart, A. A. M., Voskuil, D. W., Schreiber, G. J., Peterse, J. L., Roberts, C., Marton, M. J., *et al.* (2002). A gene-expression signature as a predictor of survival in breast cancer. *N. Engl. J. Med.*, **347**, 1999–2009.

Wan, J. C. M., Massie, C., Garcia-Corbacho, J., Mouliere, F., Brenton, J. D., Caldas, C., Pacey, S., Baird, R., and Rosenfeld, N. (2017). Liquid biopsies come of age: Towards implementation of circulating tumour DNA. *Nat. Rev. Cancer*, **17**, 223–238.

Williams, J., Kowanetz, M., Koeppen, H., Boyd, Z., Kadel, E. E., Smith, D., McCleland, M., Zou, W., and Hegde, P. S. (2016). The SP142 PD-L1 IHC assay for atezolizumab (atezo) reflects pre-existing immune status in NSCLC and correlates with PD-L1 mRNA. *Ann. Oncol.*, **27**.

Wu, A. R., Neff, N. F., Kalisky, T., Dalerba, P., Treutlein, B., Rothenberg, M. E., Mburu, F. M., Mantalas, G. L., Sim, S., Clarke, M. F., *et al.* (2014). Quantitative assessment of single-cell RNA-sequencing methods. *Nat. Methods*, **11**, 41–46.

Zhao, S. (2014). Clinical applications of next generation sequencing. *J. Health Med. Inform.*, **5**.

Chapter 7

Applications of Tissue Phenomics

Johannes Zimmermann, Nathalie Harder, and Brian Laffin
Definiens AG, Bernhard-Wicki-Strasse 5, 80636 Munich, Germany
jzimmermann@definiens.com

Recently, the investigation of the immune landscape of human cancers has made abundant scientific progress. Hallmarks of cancer such as avoidance of immune destruction and tumor-promoting inflammation are rooted in the tumor environment, which holds enormous prognostic and predictive impact (Bethmann *et al.*, 2017). Study of the immune landscape and tumor microenvironment (TME) is at the same time a paradigmatic application field for tissue phenomics, since it requires phenotyping at various scales, from individual cells to group interactions and regional measurements. Individual immuno-oncological players are characterized by protein expression levels and morphological traits, while spatial interactions and tumor geography weave in at higher orders. All of these act together to provide crucial information on immune status, function, and ultimately treatment selection.

We discuss a broad range of tissue phenomics use cases, including identification of scoring paradigms and signatures based

---

Tissue Phenomics: Profiling Cancer Patients for Treatment Decisions
Edited by Gerd Binnig, Ralf Huss, and Günter Schmidt
Copyright © 2018 Pan Stanford Publishing Pte. Ltd.
ISBN 978-981-4774-88-8 (Hardcover), 978-1-351-13427-9 (eBook)
www.panstanford.com

on hypotheses derived from basic immune oncology research. Other approaches discussed provide comprehensive visual and statistical assessment of the inflammatory TME and allow quantitative measurements of heterogeneous immune cell distribution. Finally, we examine a radically different approach, largely hypothesis-free determination of prognostic descriptors by applying an unsupervised machine learning approach.

7.1 Introduction

Tissue phenomics is the systematic discovery of quantitative descriptors for functional, morphological, and spatial patterns in histological sections that correlate with disease progression. There are a few disciplines in which this approach is playing its strengths better than the field of immune oncology, where complex interactions of various cell populations have to be understood, not least to understand immune-related adverse events by identifying immune players in specific spatial compartments such as surface epithelium, lamina propria, and intraepithelial region (Bavi *et al.*, 2017).

The tissue phenomics approach facilitates the rapid testing of hypotheses based on recent insights and current status of research. At the start of the phene (a feature or group of features in an image used to stratify or classify images or patients) discovery pipeline, a large number of phene candidates are extracted from virtual histological slides using Cognition Network Technology (CNT) (see Chapter 3), an advanced image analysis approach that emulates the human cognition process (Baatz *et al.*, 2009). These candidates— often subvisual features that are difficult or impossible for the human eye to extract—are funneled into a systematic process of evaluating their stratification potential, until top-ranking signatures are identified, which can also be combinations of readouts, representing, for example, the interplay of agents for tumor immune escape and immunogenicity, respectively. This dynamic process can be supplemented by a direct loop back into the stratum of image analysis at the beginning of the pipeline to carve out even better suited structures as carriers of the phenes in question.

7.2 Hypothesis-Driven Approaches

7.2.1 TME as a Battlefield: CD8 and PD-L1 Densities Facilitate Patient Stratification for Durvalumab Therapy in Non-Small Cell Lung Cancer

Durvalumab is an anti-PD-L1 monoclonal antibody being investigated as an immunotherapy for multiple cancer types. As with any other immunotherapy, the rate of responders is not very high, and effective biomarkers are urgently needed. The conventional patient stratification method for anti-PD-L1 therapies is the visual assessment of PD-L1 expression on tumor or tumor-infiltrating cells. A recent study showed that the product of the density of not only PD-L1(+), but also CD8(+) cells, as determined by CNT-based automated image analysis, identified patients who had significantly higher overall response rate, overall survival and progression free survival (Althammer *et al.*, 2016). Notably, this image analysis approach significantly outperformed pathologist assessment in retrospective analyses. Since the combination of PD-L1 and CD8 analyses was able to identify durva responders more effectively than either biomarker alone, it stands to reason that selecting patients for combination therapies or even determining which combination of many is the optimal choice for a given patient will grow increasingly complex. CNT-based image analysis provides not only a platform to measure the interrelationship of multiple complex biomarkers, but through the consistency in scoring and measurement inherent to image analysis approaches, insights from single agent trials can be leveraged to inform patient selection for future combination approaches. As additional indications are studied in a similar manner, it will be interesting to see which features are generalizable and which are tissue specific in terms of patient selection, and if a picture emerges not only of patients who respond favorably to checkpoint blockade but patients refractory to immunotherapy.

7.2.2 Gland Morphology and TAM Distribution Patterns Outperform Gleason Score in Prostate Cancer

In prostate cancer, we employed such a systematic discovery process and found novel biomarkers based on contextual cell density measurements, while integrating morphological information of complex metastructures (Harder *et al.*, 2017). In particular, the immune landscape in tumor-specific regions of interest (ROIs) was analyzed using dual stains related to macrophages (CD68/CD163) and T cells (CD3/CD8) as well as stains providing structural information on tissue vascularization (CD34) and glandular structures (CK18/p63, Fig. 7.1a). To determine relevant ROIs, prostate glands were categorized into tumor and healthy glands and the TME was defined as the close neighborhood of tumor gland regions (see Fig. 7.1b, orange). Immune cell–related features, including cell densities, ratios, and distances, were extracted from all ROIs (tumor and healthy glands, TME, stroma, Fig. 7.1c).

Additionally, the gland morphology and distribution were quantified using two-dimensional histograms, that is, co-occurrence matrices of different gland types (Harder *et al.*, 2016). By correlating the extracted features with the known clinical outcome of patients after radical prostectomy, a group of potential phenes providing prognostic value was identified. It turned out that a low ratio of CD8(+) cytotoxic T cells to vessel density in the tight TME was correlated with tumor progression as well as a small average distance of CD68(+) macrophages to vessels in the cancer gland regions. When considering only the gland morphology and distribution, the study showed that the mixing pattern of different sizes of healthy and cancerous glands provides prognostic value for predicting tumor progression. Validation of the discovered phene candidates on additional data acquired at different sites is required as the next step and will further drive the development of prognostic and predictive tests from these findings.

7.2.3 Novel Spatial Features Improve Staging in Colorectal Cancer

Quantitative pathology approaches can improve significantly the prognostic value of TME-related features, like in the substratification of high-risk colorectal cancer cases building on the assessment of lymphovascular invasion, tumor budding, and nuclear grade. Interobserver variability and lack of standardization prevented these features to complement the Duke classification scheme on a routine basis. However, quantification using image analysis contrib-

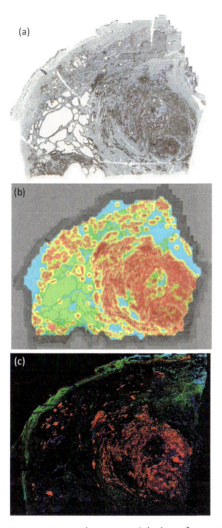

Figure 7.1 Heatmaps aggregating essential data for understanding the interactions of decisive immune players. The transfer of complex and granular cellular information to lower spatial resolution helps to carve out contextual information. (a) Original image of CK18/p63-stained tissue section. (b) ROI heatmap with tumor gland regions (red), tight TME (dark orange), wider TME (bright orange), healthy gland regions (green), stroma regions (light blue), and background or discarded border (gray). (c) Hyperspectral heatmap with six channels: average distance CD163(+) to CD68(+) macrophages (red) and vice versa (violet), CD34(+) vessel density (green), average distance of vessels to CD163(+) macrophages (blue), average distance of vessels to CD68(+) macrophages (green), tumor glands (area percentage, red).

uted the necessary consistency to suggest the use of this methodology for clinical practice (Caie *et al.*, 2016). Image analysis and big data approaches were leveraged in this instance not only to find the features that could categorize patients as low risk or high risk for disease-specific death, but also to optimize the cut-point that determines which group borderline patients should be placed in. Of particular interest in the novel features identified in this study was a measurement of the area of *poorly differentiated clusters* in the tumor sections. *Poorly differentiated* would be a subjective call from a human observer, and the area would be difficult, if not impossible, to assess in combination with this judgment call. The image analysis approach employed puts in place a reproducible, quantitative measure that can be objectively examined in light of clinical data, in the end generating a novel prognostic index that outperformed current clinical staging approaches. Providing this quantitative data also disambiguates *poorly differentiated* in a way that could speed up and augment the ability of pathologists to recognize such regions, as well as standardizing what was previously a subjective call based on experience. While in many cases the focus of advanced image analysis oncology studies is related to the drug development process, studies such as this also highlight the promise of digital pathology to enhance patient care at the level of access to state-of-the-art diagnosis and prognosis. Had algorithms such as this developed to the point that they could mimic the best human pathologists in terms of accurate clinical staging, the quality of cancer care worldwide would have advanced dramatically. Furthermore, as algorithms saw more and more images and more follow-up data were acquired, it is possible that significant improvements in clinical staging could be seen. Recent work from the same group shows that advanced nonstandard point process methods can predict colon cancer patient survival only based on unusual spatial arrangements of neoplastic nuclei (Jones-Todd *et al.*, 2017).

7.2.4 Immunoscore Is a Novel Predictor of Patient Survival in Colorectal Cancer

The Immunoscore®, a groundbreaking diagnostic tool directly linked to the TME, is the best predictor of survival in colorectal cancer patients and has the potential to guide treatment approaches for colorectal cancer (Mlecnik *et al.*, 2016). Densities of various two populations (CD3/CD8, and either CD3/CD45RO or CD8/CD45RO) are quantified by Definiens-based image analysis solutions in the center of the tumor as well as at the invasive margin and then distilled into a score ranging from I0 (low densities of both populations in both regions) to I4 (high densities of both populations in both regions). The critical insight that makes the Immunoscore successful is combining quantification of immune cell densities with regional information, which is now a centerpiece of innumerable biomarker strategies and retrospective clinical trial analyses. Immunoscore has outperformed traditional TNM staging in every study where they have been compared and has been validated in a multi-center trial (23 centers in 17 countries) by the Society for Immunotherapy in Cancer (SITC). In colorectal cancer, Immunoscore also predicts disease-specific recurrence and survival more effectively than microsatellite instability and can predict the propensity of a tumor to metastasize as well (Mlecnik *et al.*, 2016). Immunoscore has also been demonstrated to have similar predictive power in rectal tumors, and it would not be surprising to see many more indications follow suit. Yet with all the prognostic power it contains, the Immunoscore measures only two populations out of a multitude of potentially relevant cell types and, thus, is likely just the tip of the immunoprofiling iceberg. With multiplexed analysis of tissue biomarkers growing in sophistication and becoming more routine, even more powerful and informative tests could take shape. Incorporation of gene expression analysis with highly multiplexed immunofluorescence or imaging mass cytometry might provide enough histological bandwidth to not only assess all of the major immune cell populations, but also immune polarization, cytokine balance, and the immunotherapeutic target profile of a patient. This level of information will initially represent a daunting interpretive challenge, but with application of data mining and machine learning techniques, it could revolutionize personalized cancer care.

7.2.5 Analysis of Spatial Interaction Patterns Improves Conventional Cell Density Scores in Breast Cancer

The concept of immunosurveillance (Burnet, 1957) inspired a study exploring the significance of immune cell proximity to tumor cells, satisfying the increasing demand for improved and standardized scoring to exploit the prognostic potential of infiltrating immune cells (Krüger *et al.*, 2012). Cell abundance, distance metrics, neighborhood relationships, and sample heterogeneity were integrated into comprehensive assessment of immune infiltrates. Lymphocyte and macrophage subpopulations were detected by duplex immunohistochemistry (IHC) for CD3/perforin and CD68/CD163 in samples of invasive breast cancer and subsequent automated, quantitative image analysis using CNT. Recurrent infiltration patterns reflecting different grades of direct interaction between tumor and immune effector cells were identified.

When—far beyond the mere quantification of global density of lymphocytic infiltrates—the actual proximity of lymphocytes to neoplastic cells, and hence the probability of lymphocytes to actually encounter a neoplastic cell, was scrutinized, advanced CNT-based image analysis revealed a surprising behavior: In dense lymphocyte areas, significant portions of the tissue exhibited high distances from effector to target cells, whereas in cases with rather low lymphocyte densities, immune effector cells were in direct contact with the tumor cells.

This observation suggests that although effector T-cell density can appear high, infiltration is poor, lymphocytes remain in surveillance mode, and significant areas of the TME can be considered a *safe haven* for tumor cells. Breast cancer cases with much lower cell counts, however, can contain zones of intense combat.

The described methodology improved the conventional evaluation of immune cell density scores by translating objective distance metrics into reproducible, largely observer-independent interaction patterns. By incorporating immunological analysis of non-transformed breast epithelium, other studies have shed light on the baseline immune state and its natural variations, as well as non-cancerous inflammatory processes (Alfonso *et al.*, 2016; Schaadt *et al.*, 2017). Understanding immune status in this manner will not only help shape personalized medicine approaches in the immuno-oncology space, but through precise characterization of immune responses to minor tissue damage events, detection of early malignant and pre-malignant changes might be enhanced.

7.2.6 Immune Cell Infiltration is Prognostic in Breast Cancer

A series of pioneering papers by Markowetz group illustrate the power of the tissue phenomics approach even outside the integrated software environment of Definiens. Combining image analysis, gene copy number analysis, and gene expression analyses with data mining techniques, this group showed not only that image analysis could discover novel prognostic factors that would be invisible to molecular techniques, but that combining image analysis with other experimental techniques outperformed single analysis types. In addition to discovering prognostic factors based on image analysis information, these papers highlight the ability of image analysis to enhance and clarify nucleic acid–based techniques by normalizing data to tumor content and, in high grade serous ovarian cancers, demonstrated that accurate assessment of PTEN loss requires consideration of tumors stromal content (Martins *et al.*, 2014). Stromal analysis was also central to the novel prognostic features identified in ER(-) breast cancers and was discovered on an H&E-stained section (Yuan *et al.*, 2012), illustrating the unique capacity of image analysis to systematically derive tissue-based information absent any biomarker content. Morphologic and architectural analyses such as these can be embedded into analysis rulesets designed to assess biomarker content and distribution, potentially adding interpretive power to prognostic and patient selection approaches. Any tumor-promoting gene expression change, mutation, epigenetic change, or pathway activation must have a physical manifestation in the behavior of the tumor. Therefore, huge amounts of clinically relevant information likely remain undiscovered in tissue morphology and architecture, awaiting an analysis approach with sufficient power to ask the right questions and a dataset with the right clinical characteristics. Integrated approaches such as tissue phenomics and that pursued by Markowetz lab have only begun to realize their potential and will be enhanced and accelerated through the use of machine learning techniques.

7.2.7 The Immune Landscape Structure Directly Corresponds with Clinical Outcome in Clear-Cell Renal Cell Carcinoma

A recent approach used imaging mass cytometry to survey an entire cancer indication, clear-cell renal cell carcinoma (ccRCC), characterize the types of TME present, and identify potential biomarkers and targets for immunotherapy (Chevrier *et al.*, 2017). The method combines atomic mass spectrometry and flow cytometry in high spatial resolution on histological specimens. A panel of 39 markers included a number of canonical T and B cell markers, like CD8, CD4, and CD20, immunomodulatory molecules like PD-1, TIM-3, CTLA-4, and LAG-3, but also a number of newly designed macrophage markers for differentiating in detail the macrophage subsets in the samples. A key learning was that the structure of the immune landscape directly allowed to stratify patients by applying a correspondence analysis that captures the complexity of relationships between immune cell species within the TME. Other findings showed that while PD-1 is broadly expressed across T-cell species, other important targets such as CTLA-4 are much lower expressed and might, therefore, be less effective in immunotherapy, and CD38, known as marker of T-cell exhaustion in infectious diseases, might also play a crucial role in ccRCC.

7.3 Hypothesis-Free Prediction

Beck *et al.* (2011) used an unbiased data-driven approach to discover prognostically significant morphologic features in breast cancer. This discovery-based approach has been widely used in the analysis of genomic data, but not yet in the study of cancer morphology from microscopic images of patient samples. Microscopic images of cancer samples represent a rich source of biological information, because this level of resolution facilitates the detailed quantitative assessment of cancer cells' relationships with each other, with normal cells, and with the TME, all of which represent key *hallmarks of cancer*. They developed a customized image processing pipeline within the Definiens datafication environment to measure a rich quantitative set of more than 6000 features from the breast cancer epithelium and stroma. Those included standard morphometric descriptors of image objects as well as higher-level contextual, relational, and global attributes. These measurements were used to construct a prognostic model, which was applied to microscopic images from two independent cohorts of breast cancer patients. The prognostic model score was strongly associated with overall survival in both cohorts, independent of clinical, pathological, and molecular factors. The image analysis system was automated without any manual steps, greatly increasing its scalability. In addition, the system measured thousands of morphologic descriptors of diverse elements of the microscopic cancer image allowing identification of prognostic features: The system revealed three stromal features to be significantly associated with survival, which were previously unrecognized prognostic determinants for breast cancer. Beck *et al.* concluded that the unbiased method and flexible architecture of the system might be used for building a library of image-based models in multiple cancer types, each optimized to predict a specific clinical outcome, including response to particular pharmacologic agents, allowing this approach to be used to guide treatment decisions. This study was considered to be the first truly objective, quantitative grading system for cancerous tissue and surrounding stroma. The use of spatial information contributes a dimension that traditional molecular analysis of homogenized tissue will never be able to provide: "Remember, from the whole genome sequence perspective, there is no difference between a caterpillar and a butterfly" (Rimm, 2011).

7.4 Summary

Definiens technology, which has been widely used in the field of immune oncology (Table 7.1), enables a broad range of biomarker studies in multiple cancer types and finds its greatest utility when image analysis is fed into a big data type approach, which is the heart of the tissue phenomics paradigm. Its unique potential can be fully leveraged if complex relational questions are interlocked with functional and morphological analyses, as is frequently the case in large clinical studies in immune oncology.

Table 7.1 Selected examples for analysis of immuno-oncologically relevant markers using Definiens CNT (IHC: immunohistochemistry, IF: immunofluorescence)

Biomarker	Modality	Indication	Reference
CD3	IHC	Solid tumor immunotherapy	(Stadler *et al.*, 2017)
FOXP3, CD3, CD8	IHC	Melanoma	(Sahin *et al.*, 2017)
FOXP3, CD4, CD8	IF	Melanoma	(Kreiter *et al.*, 2015)
FOXP3, CD4, CD8	IHC	Prostate cancer	(Linch *et al.*, 2017)
CD8	IHC	Bladder cancer	(Rosenberg *et al.*, 2016)
CD8, CD68, CD163, IDO	IHC	Sarcomas, GIST	(Toulmonde *et al.*, 2017)
FOXP3, CD4, CD8, CD20	IHC	NSCLC	(Kinoshita *et al.*, 2016)
FOXP3, CD8	IHC	Colorectal cancer	(Saito *et al.*, 2016)
CD204	IHC	Gastric cancer	(Ichimura *et al.*, 2016)
CD204	IHC	Upper urinary tract cancer	(Ichimura *et al.*, 2014)
CD3, CD8, CD20, CD45RO	IHC	Upper urinary tract cancer	(Makise *et al.*, 2015)
CD3, CD4, CD8, FOXP3, Cbl-b, BTLA	IHC	Gall bladder cancer	(Oguro *et al.*, 2015)
KI67, CD3, CD20	IF	Colorectal cancer	(Mlecnik *et al.*, 2016)
CD3, CD8	IHC	Rectal cancer	(Anitei *et al.*, 2014)
KI67, F4/80, MMP-9, CSF1R	IHC	Prostate cancer	(Escamilla *et al.*, 2015)
FOXP3, CD68, CD34	IHC	Diffuse large B cell lymphoma	(Gomez-Gelvez *et al.*, 2016)
p-eIF2α	IHC	Ductal carcinoma *in situ*	(Semeraro *et al.*, 2016)
CD8, CLA	IF	Merkel cell carcinoma	(Afanasiev *et al.*, 2013)

Biomarker	Modality	Indication	Reference
KI67, CD3, CD20	IF	Various indications	(Mlecnik *et al.*, 2014)
CD3, CD8	IHC	Metastasis	(Berghoff *et al.*, 2016)
CD20, DC-LAMP	IHC	Metastasis	(Montfort *et al.*, 2016)
Tumor area, micrometastases, NKp46, NK1.1	H&E, IHC	Metastasis	(Paolino *et al.*, 2014)

Tissue phenomics not only helps to elucidate mechanisms of action or shape upcoming immune therapy concepts but also can integrate multiple data sources into a seamless whole, magnifying the power of the experimental techniques employed with additional levels of context and normalization not available to the individual approaches. As combination therapies become more commonplace in clinical settings, the complexity of the decision-making environment for physicians will increase in tandem. Therefore, the relevance of the integrative analysis capacity of tissue phenomics will also grow, as it provides a rapid and systematic method for sifting through masses of data for relevant correlations to clinical outcome, and for continuously integrating new information. While image analysis will provide critical contextual information in any system designed to select patients for a broad range of therapies, physicians are unlikely to be willing or able to order dozens of IHC tests due to cost. Thus, we propose that tissue phenomics represents the most efficient approach currently available for creation and interpretation of such a patient selection system, which could be broadened to generate a comprehensive decision support system for physicians soon required to cope with mountains of data and guidelines related to dozens of viable therapeutic approaches. Incorporation of the morphological features discovered in the context of predicting clinical outcome from H&E sections would provide additional power to such a system, as it would inform how aggressively a patient should be treated from the array of therapeutic approaches that they were well suited for. Application of consistent image analysis methods across multiple studies and therapeutic targets could help streamline the decision-making process further, through the identification of features associated with treatment failure, resistance, and recurrence in both IO and traditional oncology settings.

References

Afanasiev, O. K., Nagase, K., Simonson, W., Vandeven, N., Blom, A., Koelle, D. M., Clark, R., and Nghiem, P. (2013). Vascular E-selectin expression correlates with CD8 lymphocyte infiltration and improved outcome in Merkel cell carcinoma. *J. Invest. Dermatol.*, **133**, 2065–2073.

Alfonso, J. C. L., Schaadt, N. S., Schönmeyer, R., Brieu, N., Forestier, G., Wemmert, C., Feuerhake, F., and Hatzikirou, H. (2016). In-silico insights on the prognostic potential of immune cell infiltration patterns in the breast lobular epithelium. *Sci. Rep.*, **6**, 33322.

Althammer, S., Steele, K., Rebelatto, M., Tan, T. H., Wiestler, T., Schmidt, G., Higgs, B., Li, X., Shi, L., Jin, X., *et al.* (2016). Combinatorial CD8+ and PD-L1+ cell densities correlate with response and improved survival in non-small cell lung cancer (NSCLC) patients treated with durvalumab. *J. Immunother. Cancer*, **4**, 191.

Anitei, M.-G., Zeitoun, G., Mlecnik, B., Marliot, F., Haicheur, N., Todosi, A.-M., Kirilovsky, A., Lagorce, C., Bindea, G., Ferariu, D., *et al.* (2014). Prognostic and predictive values of the immunoscore in patients with rectal cancer. *Clin. Cancer Res.*, **20**, 1891–1899.

Baatz, M., Zimmermann, J., and Blackmore, C. G. (2009). Automated analysis and detailed quantification of biomedical images using Definiens Cognition Network Technology. *Comb. Chem. High Throughput Screen.*, **12**, 908–916.

Bavi, P., Butler, M., Serra, S., and Chetty, R. (2017). Immune modulator-induced changes in the gastrointestinal tract. *Histopathology*, **71**, 494–496.

Beck, A. H., Sangoi, A. R., Leung, S., Marinelli, R. J., Nielsen, T. O., van de Vijver, M. J., West, R. B., van de Rijn, M., and Koller, D. (2011). Systematic analysis of breast cancer morphology uncovers stromal features associated with survival. *Sci. Transl. Med.*, **3**, 108ra113.

Berghoff, A. S., Fuchs, E., Ricken, G., Mlecnik, B., Bindea, G., Spanberger, T., Hackl, M., Widhalm, G., Dieckmann, K., Prayer, D., *et al.* (2016). Density of tumor-infiltrating lymphocytes correlates with extent of brain edema and overall survival time in patients with brain metastases. *Oncoimmunology*, **5**, e1057388.

Bethmann, D., Feng, Z., and Fox, B. A. (2017). Immunoprofiling as a predictor of patient's response to cancer therapy: Promises and challenges. *Curr. Opin. Immunol.*, **45**, 60–72.

Burnet, M. (1957). Cancer: A biological approach. *Br. Med. J.*, **1**, 779–786.

Caie, P. D., Zhou, Y., Turnbull, A. K., Oniscu, A., and Harrison, D. J. (2016). Novel histopathologic feature identified through image analysis augments stage II colorectal cancer clinical reporting. *Oncotarget*, **7**, 44381–44394.

Chevrier, S., Levine, J. H., Zanotelli, V. R. T., Silina, K., Schulz, D., Bacac, M., Ries, C. H., Ailles, L., Jewett, M. A. S., Moch, H., *et al.* (2017). An immune atlas of clear cell renal cell carcinoma. *Cell*, **169**, 736–749.e18.

Escamilla, J., Schokrpur, S., Liu, C., Priceman, S. J., Moughon, D., Jiang, Z., Pouliot, F., Magyar, C., Sung, J. L., Xu, J., *et al.* (2015). CSF1 receptor targeting in prostate cancer reverses macrophage-mediated resistance to androgen blockade therapy. *Cancer Res.*, **75**, 950–962.

Gomez-Gelvez, J. C., Salama, M. E., Perkins, S. L., Leavitt, M., and Inamdar, K. V. (2016). Prognostic impact of tumor microenvironment in diffuse large B-cell lymphoma uniformly treated with R-CHOP chemotherapy. *Am. J. Clin. Pathol.*, **145**, 514–523.

Harder, N., Athelogou, M., Hessel, H., Buchner, A., Schönmeyer, R., Schmidt, G., Stief, C. G., Kirchner, T., and Binnig, G. (2016). Co-occurrence features characterizing gland distribution patterns as new prognostic markers in prostate cancer whole-slide images. In *2016 IEEE 13th International Symposium on Biomedical Imaging* (*ISBI*), pp. 807–810.

Harder, N., Athelogou, M., Hessel, H., Brieu, N., Yigitsoy, M., Zimmermann, J., Baatz, M., Buchner, A., Stief, C. G., Kirchner, T., *et al.* (2017). Tissue Phenomics for prognostic biomarker discovery in low- and intermediate-risk prostate cancer. *Mod. Pathol.*, *Submitted for publication.*

Ichimura, T., Morikawa, T., Kawai, T., Nakagawa, T., Matsushita, H., Kakimi, K., Kume, H., Ishikawa, S., Homma, Y., and Fukayama, M. (2014). Prognostic significance of CD204-positive macrophages in upper urinary tract cancer. *Ann. Surg. Oncol.*, **21**, 2105–2112.

Ichimura, T., Abe, H., Morikawa, T., Yamashita, H., Ishikawa, S., Ushiku, T., Seto, Y., and Fukayama, M. (2016). Low density of CD204-positive M2 type tumor-associated macrophages in Epstein–Barr virus-associated gastric cancer: A clinicopathological study with digital image analysis. *Hum. Pathol.*, **56**, 74–80.

Jones-Todd, C. M., Caie, P., Illian, J., Stevenson, B. C., Savage, A., Harrison, D. J., and Bown, J. L. (2017). Unusual structures inherent in point pattern data predict colon cancer patient survival. *Ann. Appl. Stat.*, *submitted for publication*.

Kinoshita, T., Muramatsu, R., Fujita, T., Nagumo, H., Sakurai, T., Noji, S., Takahata, E., Yaguchi, T., Tsukamoto, N., Kudo-Saito, C., *et al.* (2016). Prognostic value of tumor-infiltrating lymphocytes differs depending on histological type and smoking habit in completely resected non-small-cell lung cancer. *Ann. Oncol.*, **27**, 2117–2123.

Kreiter, S., Vormehr, M., van de Roemer, N., Diken, M., Löwer, M., Diekmann, J., Boegel, S., Schrörs, B., Vascotto, F., Castle, J. C., *et al.* (2015). Mutant MHC class II epitopes drive therapeutic immune responses to cancer. *Nature*, **520**, 692–696.

Krüger, J. M., Wemmert, C., Sternberger, L., Bonnas, C., Dietmann, G., Gançarski, P., and Feuerhake, F. (2012). Combat or surveillance? Evaluation of the heterogeneous inflammatory breast cancer microenvironment. *J. Pathol.*, **229**, 569–578.

Linch, M., Goh, G., Hiley, C., Shanmugabavan, Y., McGranahan, N., Rowan, A., Wong, Y. N. S., King, H., Furness, A., Freeman, A., *et al.* (2017). Intratumoural evolutionary landscape of high risk prostate cancer: The PROGENY study of genomic and immune parameters. *Ann. Oncol.*, **28**, 2472–2480.

Makise, N., Morikawa, T., Nakagawa, T., Ichimura, T., Kawai, T., Matsushita, H., Kakimi, K., Kume, H., Homma, Y., and Fukayama, M. (2015). MAGE-A expression, immune microenvironment, and prognosis in upper urinary tract carcinoma. *Hum. Pathol.*, **50**, 62–69.

Martins, F. C., Santiago, I. de, Trinh, A., Xian, J., Guo, A., Sayal, K., Jimenez-Linan, M., Deen, S., Driver, K., Mack, M., *et al.* (2014). Combined image and genomic analysis of high-grade serous ovarian cancer reveals PTEN loss as a common driver event and prognostic classifier. *Genome Biol.*, **15**, 526.

Mlecnik, B., Bindea, G., Angell, H. K., Sasso, M. S., Obenauf, A. C., Fredriksen, T., Lafontaine, L., Bilocq, A. M., Kirilovsky, A., Tosolini, M., *et al.* (2014). Functional network pipeline reveals genetic determinants associated with *in situ* lymphocyte proliferation and survival of cancer patients. *Sci. Transl. Med.*, **6**, 228ra37.

Mlecnik, B., Bindea, G., Angell, H. K., Maby, P., Angelova, M., Tougeron, D., Church, S. E., Lafontaine, L., Fischer, M., Fredriksen, T., *et al.* (2016). Integrative analyses of colorectal cancer show immunoscore is a stronger predictor of patient survival than microsatellite instability. *Immunity*, **44**, 698–711.

Montfort, A., Pearce, O. M. T., Maniati, E., Vincent, B., Bixby, L. M., Böhm, S., Dowe, T., Wilkes, E. H., Chakravarty, P., Thompson, R., *et al.* (2016). A strong B cell response is part of the immune landscape in human high-grade serous ovarian metastases. *Clin. Cancer Res.*, **23**, 250–262.

Oguro, S., Ino, Y., Shimada, K., Hatanaka, Y., Matsuno, Y., Esaki, M., Nara, S., Kishi, Y., Kosuge, T., and Hiraoka, N. (2015). Clinical significance of tumor-infiltrating immune cells focusing on BTLA and Cbl-b in patients with gallbladder cancer. *Cancer Sci.*, **106**, 1750–1760.

Paolino, M., Choidas, A., Wallner, S., Pranjic, B., Uribesalgo, I., Loeser, S., Jamieson, A. M., Langdon, W. Y., Ikeda, F., Fededa, J. P., *et al.* (2014). The E3 ligase Cbl-b and TAM receptors regulate cancer metastasis via natural killer cells. *Nature*, **507**, 508–512.

Rimm, D. L. (2011). C-path: A Watson-like visit to the pathology lab. *Sci. Transl. Med.*, *3*, 108fs8.

Rosenberg, J. E., Hoffman-Censits, J., Powles, T., van der Heijden, M. S., Balar, A. V., Necchi, A., Dawson, N., O'Donnell, P. H., Balmanoukian, A., Loriot, Y., *et al.* (2016). Atezolizumab in patients with locally advanced and metastatic urothelial carcinoma who have progressed following treatment with platinum-based chemotherapy: A single-arm, multicentre, phase 2 trial. *Lancet Lond. Engl.*, **387**, 1909–1920.

Sahin, U., Derhovanessian, E., Miller, M., Kloke, B.-P., Simon, P., Löwer, M., Bukur, V., Tadmor, A. D., Luxemburger, U., Schrörs, B., *et al.* (2017). Personalized RNA mutanome vaccines mobilize poly-specific therapeutic immunity against cancer. *Nature*, advance online publication.

Saito, T., Nishikawa, H., Wada, H., Nagano, Y., Sugiyama, D., Atarashi, K., Maeda, Y., Hamaguchi, M., Ohkura, N., Sato, E., *et al.* (2016). Two FOXP3(+) CD4(+) T cell subpopulations distinctly control the prognosis of colorectal cancers. *Nat. Med.*, **22**, 679–684.

Schaadt, N. S., Alfonso, J. C. L., Schönmeyer, R., Grote, A., Forestier, G., Wemmert, C., Krönke, N., Stoeckelhuber, M., Kreipe, H. H., Hatzikirou, H., *et al.* (2017). Image analysis of immune cell patterns in the human mammary gland during the menstrual cycle refines lymphocytic lobulitis. *Breast Cancer Res. Treat.*, **164**, 305–315.

Semeraro, M., Adam, J., Stoll, G., Louvet, E., Chaba, K., Poirier-Colame, V., Sauvat, A., Senovilla, L., Vacchelli, E., Bloy, N., *et al.* (2016). The ratio of CD8+/FOXP3 T lymphocytes infiltrating breast tissues predicts the relapse of ductal carcinoma *in situ*. *Oncoimmunology*, **5**, e1218106.

Stadler, C. R., Bähr-Mahmud, H., Celik, L., Hebich, B., Roth, A. S., Roth, R. P., Karikó, K., Türeci, Ö., and Sahin, U. (2017). Elimination of large tumors in mice by mRNA-encoded bispecific antibodies. *Nat. Med.*, advance online publication.

Toulmonde, M., Penel, N., Adam, J., Chevreau, C., Blay, J.-Y., Le Cesne, A., Bompas, E., Piperno-Neumann, S., Cousin, S., Grellety, T., *et al.* (2017). Use of PD-1 targeting, macrophage infiltration, and IDO pathway activation in sarcomas: A phase 2 clinical trial. *JAMA Oncol*, doi: 10.1001/jamaoncol.2017.1617.

Yuan, Y., Failmezger, H., Rueda, O. M., Ali, H. R., Gräf, S., Chin, S.-F., Schwarz, R. F., Curtis, C., Dunning, M. J., Bardwell, H., *et al.* (2012). Quantitative image analysis of cellular heterogeneity in breast tumors complements genomic profiling. *Sci. Transl. Med.*, **4**, 157ra143.

Chapter 8

Tissue Phenomics for Diagnostic Pathology

Maria Athelogou and Ralf Huss
Definiens AG, Bernhard-Wicki-Strasse 5, 80636 Munich, Germany
mathelogou@definiens.com

Tissue phenomics is a key component for the successful transition of today's digital pathology into the next-generation diagnostic pathology (Pathology 4.0). The intelligent combination of integrated workflows with effective user and application interfaces, cognition networks, and hypothesis-free machine learning algorithms based on the conventional expert pathology wisdom will generate new knowledge and diagnostic innovation in the routine diagnostic pathology and tissue biomarker discovery.

8.1 Introduction

The major goal of automated image and data analysis in digital pathology is the development of algorithms and methods for clinical applications and advanced diagnostics. Improvements in

---

Tissue Phenomics: Profiling Cancer Patients for Treatment Decisions
Edited by Gerd Binnig, Ralf Huss, and Günter Schmidt
Copyright © 2018 Pan Stanford Publishing Pte. Ltd.
ISBN 978-981-4774-88-8 (Hardcover), 978-1-351-13427-9 (eBook)
www.panstanford.com

data acquisition and management technologies induced the need to deal with *big data* also in the clinical environment. New big data approaches combined with current developments in disciplines such as artificial intelligence and machine learning have led to the discovery of new and highly advanced tissue-based biomarkers, complex methodologies, and broad applications in digital pathology.

The primary goal of image and data analysis for diagnostic purposes is to increase the amount and quality of data that are derived from a pathological specimen by adding quantitative measurements to histologic features in representative regions (regions of interest). Digital image analysis offers pathologists the opportunity to broaden their diagnostic capabilities and to transition anatomical pathology from a subjective and mostly qualitative diagnosis toward an objective, quantitative, and robust diagnostic platform based on measurable end points using computer-assisted signatures (Webster and Dunstan, 2014).

8.2   Digital Pathology

Within the past decade, several image analysis–based diagnostic tests on tissue have received 510(K) clearance from the US Food and Drug Administration (FDA) for diagnostic use, for example, in breast cancer. However, image analysis is being applied increasingly to many applications in an exploratory setting, including the assessment of, for example, tissue biobank quality assurance, automated scoring of animal models (toxicology and non-clinical safety), tissue microarray construction, assessments of protein expression levels in tissues and cell-based microarrays.

The diagnosis of a disease or condition is composed of several datasets comprising (a) clinical history of the patient and development/progression of the disease (medical history), (b) blood and serum analysis (clinical pathology), (c) screening for localized or generalized conspicuous foci in tissues or organs (imaging and radiomics), (d) structural changes within such lesions after biopsy or resection at the microscopic level (anatomical pathology), (e) functional data such as receptor expression and genetic alterations (surgical pathology) (Kayser *et al.*, 2015).

Anatomical pathology comprises all diagnostic tools and algorithms that serve diagnostic purposes from tissue, cells, or liquids. In *surgical pathology*, the *findings* may include also data that are derived from (a) image content information, (b) clinical history, (c) expertise of the pathologist, and (d) pre-existing knowledge about the disease. *Digital pathology* uses such anatomical and surgical pathology approaches together with statistical and decision algorithms (neural networks, discriminate analysis, factor analysis, *etc.*) to identify the most probable diagnosis (Kayser *et al.*, 2015). Several categories can be distinguished, namely (a) conventional or *classical histological* and *cytological* diagnosis, (b) *prospective diagnosis*, (c) *indicative* diagnosis, and (d) *risk-assigned* diagnosis (Görtler *et al.*, 2006).

The *classical* histopathological diagnosis is the necessary prerequisite for any reliable treatment of diseases such as cancer and is, by far, the least expensive diagnostic procedure. It usually only requires empirical expert knowledge and the descriptive assessment of a stained tissue section under the microscope. Instead a prognosis-associated or indicative (predictive) information requires additional and detailed information/data. The recognition of a *risk-associated disease*, such as a genetic predisposition to develop cancer, is usually performed by highly specialized institutions or expert departments. Therefore, institutions involved in advanced tissue-based diagnostics must have access to a broad range of relevant sources of clinical data, information, and knowledge. Such advanced labs and pathology specialists can provide integrated and highly detailed information on the disease and directions toward the most promising treatment (indicative/predictive information) (Görtler *et al.*, 2006).

Extended tissue-based diagnosis also includes all diagnostic procedures that further analyze spatial relationships in all structural and biological functional units. Additional contextual information are analyzed in advanced studies, which provide the foundation of tissue phenomics.

Digital pathology tools collect relevant data from all available information sources in a standardized manner (according to the regulatory framework) and apply computational algorithms to yield a hierarchy of results associated with the disease condition, which will determine the most promising treatments or ongoing clinical trials. However, the integration and standardization of such digital pathology tools in a clinical and strictly regulated clinical environment (applying GCP regulations) seem to be one major obstacle for the immediate and sustainable implementation of digital pathology into routine pathology (Görtler *et al.*, 2006).

Meeting the challenges of the digital pathology workflow, including the validation of the digital image quality, will not compromise the diagnostic performance of anatomical and surgical pathologists reading whole-slide images on computer screens instead of traditional slide reading with the microscope (Taylor, 2014). The additional use of image quantification measures based on digital image analysis will transform the diagnostic processes in routine pathology (Kayser *et al.*, 2009).

One goal of image and data analysis is to further optimize and standardize existing scores and add additional information, which is in alignment with the visual inspection and qualitative evaluation of the tissue morphology. Histopathological tissue analysis by a pathologist represents currently only the definitive method to confirm the presence or absence of disease, to grade the tumor, and to measure the disease progression (Webster and Dunstan, 2014). For example, the current Gleason grading scheme is used to predict prostate cancer prognosis and help guide therapy. The Gleason grading system is based solely on architectural patterns. In 2005, the International Society of Urologic Pathologists together with the World Health Organization (WHO) made a series of recommendations to modify the Gleason grading system (Luthringer and Gross, 2001). Computer-aided diagnosis can add or increase accuracy of predictive diagnosis and patient management decisions regarding therapy and outcome evaluation (Athelogou *et al.*, 2014).

8.3 Tissue Phenomics Applications for Computer-Aided Diagnosis in Pathology

Several solutions for digital diagnosis have already been developed, which are based on the tissue phenomics image analysis technology and the corresponding image analysis platform. Some of them are already integrated into the clinical pathology routine.

Clarient Inc. (now part of NeoGenomics Laboratories), in cooperation with Definiens AG, developed several solutions for the automated scoring of breast cancer and colorectal cancer. Such solutions use image analysis algorithms for immunohistochemistry (IHC) assays, including biomarkers such as ER, PR, HER2, KI67, P53, EGFR, PMS2, MSH6, MLH1, MSH2, BCL1, AR, p21, and p27. The solutions support hundreds of pathologists in order to analyze biomarker expression in thousands of digital tissue slides per day with high consistency and reproducibility. These automated solutions can also be integrated into a CLIA laboratory workflow and are designed for images from all major scanners.

Metamark's ProMark® prognostic test was launched in 2014 with CMS coverage in 2016. Tissue sections are subjected to multiplex immunofluorescent (IF) and counterstained with DAPI (for nuclear detection) to be used as a proprietary assay format that enables the quantitative biomarker measurements in the region of interest on the entire tissue section. The *raw* images of the stained sections are acquired on the image acquisition platform and transferred to the imaging platform. Through advanced object recognition and the use of a proprietary script, quantitative measurements of digitized biomarkers are possible. This solution is a tool for treatment decisions in prostate cancer, adding information besides the conventional Gleason score, also to overcome tumor heterogeneity.

Based on research by Galon *et al.* (2012) and applying tissue phenomics, HalioDx Inc. developed the Immunoscore®. The Immunoscore integrates IHC in combination with advanced imaging analysis solutions to enable the extraction of spatial information in colon cancer. This is a diagnostic tool for routine pathology and exploratory application in immuno-oncology, which offers image and data analysis for the identification and quantification of certain immune cells in predefined cancer regions (*e.g.*, immune cell density in the invasive margin versus the tumor center). While the Immunoscore is currently designed for colon cancer, it has the potential to be a platform for many cancers to determine the immune landscape in individual tumors to better predict disease progression and response to therapy. The clinical evaluation is currently ongoing.

8.3.1   Technical Prerequisites for Tissue Phenomics

Vendors of glass slide scanners and image analysis software have developed solutions for digital pathology diagnosis. Leica Biosystems received FDA clearance for the Aperio eIHC, an *in vitro* diagnostic solution to assist pathologists in the interpretation of ER, PR, and HER2 breast IHC. Aperio eIHC system is indicated for use as an aid in the management, prognosis, and prediction of therapy outcomes in breast cancer, as accessory tool to score the DAKO ER, PR, and HercepTest™.

Philips IntelliSite Pathology Solution (PIPS) is currently available for primary diagnostic use in the USA. The FDA has recently permitted its marketing and sales, and this is, therefore, the first whole-slide imaging system that is approved for review and interpretation of digital surgical pathology slides prepared from biopsied tissue. This is also the first time that the FDA has approved whole-slide imaging system for this purpose. The system enables pathologists to read tissue slides digitally on a single or multiple computer screens in order to make diagnoses, rather than looking directly at a tissue sample mounted on a glass slide under a conventional light microscope.

8.4   Tissue Phenomics Applications for Decision Support Systems in Pathology

The future of tissue phenomics in digital pathology will lead to the implementation of (semi-)automated digital pathology solutions as an end-to-end diagnostic workflow. Software applications will apply machine learning solutions for automatic analysis of diagnostic slides in order to help pathologists and other experts to improve diagnostic accuracy to reduce the human error rate and to utilize image archives for comprehensive reference.

In the future, digital pathology workflows will provide tools and resources for pathologists and clinicians to effectively and accurately share information via an integrated digital *pathology cockpit* to analyze large sets of tissue slides using image analysis algorithms to create and distribute comprehensive reports for the exchange of documents and knowledge. Future digital pathology workflows will integrate slide scanners with interfaces for laboratory information systems (LIS), image analysis platforms and user interfaces for the visual assessment and applications for annotating objects and regions of interest and to support a cloud integration (browser-based solutions). Cloud integration will enable pathologist to start the process of annotations and start analysis of images by simply uploading images to the appropriate web workspace, requesting the desired analysis or automated image analysis solutions. If not available yet, imaging experts will provide novel image analysis solutions, to be added to the pathologist's web workspace on the server.

The development of image analysis applications in a cloud environment by applying machine learning and artificial intelligence methods provides many advantages for the image analysis experts as well as for the research pathologist. Machine learning algorithms learn to recognize images and patterns almost the same way as humans do, but there is no need to formalize complex *handcrafted features*. The pathologist just needs to annotate those structures, which they want to be recognized (*e.g.*, cancer regions versus the surrounding stroma). The system can be trained on large datasets to achieve a robust recognition solution with increased accuracy.

The diagnostic use of highly sophisticated algorithms and standardized methods will enable the analysis of big data but also to analyze very complex datasets to achieve a maximal exploitation of available information from these data. The optimal extraction of such information from big data in tissue sections with tissue phenomics requires an easy-to-access platform that can be used by experts from multiple domains, for example, pathologists, biomarker experts, and image analysis specialists. Co-registered multi-viewing opportunities (Schönmeyer *et al.*, 2014) and the simultaneous analysis of multiple biomarkers on single or consecutive sections (Schönmeyer *et al.*, 2015) expands the value and reliability of data generation and support the decision-making process. Tissue phenomics user interface (UI) not only allows slide/image navigation, changes of magnification, snap shots, contemporary viewing of several images (*e.g.*, H&E and IHC images synchronously), but also employs fully automated co-registration of serial sections (*virtual multiplexing*) and fully automated transfer of annotated regions from the annotated single slide into the remaining co-registered serial slides.

Fully automated tissue scoring and biomarker quantification within those annotated tissue regions or within the whole tissue image is delivered on demand.

The future development of decision support systems (DSSs) comprises algorithms and methods for computer-aided diagnosis, quality assurance, image viewing, inspection and context navigation, co-registration, virtual multiplexing, and data analysis. The development of such DSS might require the co-development of the diagnostic assays and the corresponding *complementary/companion* tests.

8.5 Summary of Pathology 4.0

After paradigm shifts in physics, biology, and psychology, the fourth revolution is mostly about the generation of new knowledge, the storage of all accessible data and their expedited exchange through broadly available information technologies. The necessary digitization and intelligent data processing will play an increasingly important role in our daily lives and particularly in the health care industry and medicine. The availability of all relevant information in patient care will dramatically improve the diagnosis of diseases and the selection of effective treatments. Tissue phenomics is an important cornerstone of Pathology 4.0, which uses digitized tissue images for intelligent image mining and comprehensive data processing to generate a wealth of new medical knowledge from pathological specimens along with novel clinical insights. New drugs or drug combinations and (complementary/companion) diagnostic tests will become available, and the *medical* error rate will eventually remain on an insignificant level.

Digitization of images that can be read on a hand-held device or a computer screen instead of a conventional microscope would not have such a disruptive impact on the future of pathology, unless it goes in parallel with intelligent data processing (including information from genomics and patient history), the availability of data networks with enormous and fast data streams and information technologies such as machine learning, automatic feature (phene) cognition, cloud computing, and big data.

Pathology has already undergone changes over the last centuries with the gradual emergence of novel technologies, initially with the use of better microscopes with higher resolutions and the development of genetic testing. In the present age of *information and communication technologies*, also the practice of pathology has started to change and to adopt to the availability of more (big) and very complex data. Most of the currently practicing pathologists have been trained through reading analog information in textbooks and scientific publications, but it seems impossible to keep up with the growing body of relevant information to properly diagnose a disease with all its complex sub-entities and molecularly defined genotypes that require sometime very different treatment modalities.

While digitization in pathology started with *telepathology*, which was the transfer of images to obtain a second or third expert opinion from or to remote places to pathology centers, it is now deploying artificial intelligence (or hybrid technologies) and machine learning to identify known and complex patterns with higher confidence (supervised) or recover novel features or phenes (unsupervised). Tissue phenomics is a tool that combines supervised and unsupervised image and data mining to implement spatial relationships in a compartmental context, for example, invasive cancer in an inflammatory stroma.

Pathology 4.0 is not only about intelligent image and data processing but also about an advanced workflow that deals with the generating and capturing of images, the documentation and management of medical content, and the distribution and transfer of knowledge and its re-use and re-purpose of *wisdom* depending on even more advanced information as it emerges. This also requires a high degree of connectivity of all stakeholders through a functional human–computer interface that also complies with the regulatory framework.

Pathology 4.0 will be an indispensable approach to transition empirical medical wisdom to an effective precision medicine tool to better stratify patients and improve the outcome of the individual disease.

Acknowledgments

The authors thank Thomas Colarusso for providing commercial examples of applications of tissue phenomics.

References

Athelogou, M., Kirchner, T., Hessel, H., and Binnig, G. (2014). Gleason grading by segmenting and combining co-registered images of differently stained tissue slices. *United States Patent 8879819 B2*.

Luthringer, D. J. and Gross, M. (2001). Gleason grade migration: Changes in prostate cancer grade in the contemporary era. *PCRI Insights*, **9**, 2–3.

Galon, J., Pagès, F., Marincola, F. M., Thurin, M., Trinchieri, G., Fox, B. A., Gajewski, T. F., and Ascierto, P. A. (2012). The immune score as a new possible approach for the classification of cancer. *J. Transl. Med.*, **10**, 1.

Görtler, J., Berghoff, M., Kayser, G., and Kayser, K. (2006). Grid technology in tissue-based diagnosis: Fundamentals and potential developments. *Diagn. Pathol.*, **1**, 23.

Kayser, K., GÅśrtler, Jă., Bogovac, M., Bogovac, A., Goldmann, T., Vollmer, E., and Kayser, G. (2009). AI (artificial intelligence) in histopathology: From image analysis to automated diagnosis. *Folia Histochem. Cytobiol.*, **47**, 355–361.

Kayser, K., Borkenfeld, S., Carvalho, R., and Kayser, G. (2015). How to implement digital pathology in tissue-based diagnosis (surgical pathology)? *Diagn. Pathol.*, **1**.

Schönmeyer, R., Athelogou, M., Schmidt, G., and Binnig, G. (2014). Visualization and navigation platform for co-registered whole tissue slides. In *Bildverarbeitung für die Medizin 2014*, T. M. Deserno, H. Handels, H.-P. Meinzer, and T. Tolxdorff, eds. (Springer Berlin Heidelberg), pp. 13–18.

Schönmeyer, R., Brieu, N., Schaadt, N., Feuerhake, F., Schmidt, G., and Binnig, G. (2015). Automated whole slide analysis of differently stained and co-registered tissue sections. In *Bildverarbeitung für die Medizin 2015*, H. Handels, T. M. Deserno, H.-P. Meinzer, and T. Tolxdorff, eds. (Springer Berlin Heidelberg), pp. 407–412.

Taylor, C. R. (2014). Issues in using whole slide imaging for diagnostic pathology: "Routine" stains, immunohistochemistry and predictive markers. *Biotech. Histochem.*, **89**, 419–423.

Webster, J. D. and Dunstan, R. W. (2014). Whole-slide imaging and automated image analysis: Considerations and opportunities in the practice of pathology. *Vet. Pathol.*, **51**, 211–223.

Chapter 9

Digital Pathology: Path into the Future

Peter D. Caie and David J. Harrison
School of Medicine, University of St Andrews, North Haugh,
St Andrews KY16 9TF, UK
pdc5@st-andrews.ac.uk

Digital pathology is now rapidly translating to the clinic, facilitating multiple advantages compared to traditional histopathology. Regulatory approval and the advance in associated technology have expedited this translation. Whole-slide scanners are now capable of standardized batched image capture coupled to a fully digital workflow, including patient records. In the near future, patients with no local access to specialist pathology resource will benefit from remote diagnoses across expert networks spanning the globe. This is even true for low-resource countries with the development of mobile and cost-effective digital pathology solutions. Automated image analysis will not be far behind in clinical translation. Algorithms can standardize the quantification of histopathological features and biomarkers while taking into account their spatial interaction within a complex tissue. Machine learning will also feature in the future of clinical histopathology as the computer learns to exclude tissue- and

---

Tissue Phenomics: Profiling Cancer Patients for Treatment Decisions
Edited by Gerd Binnig, Ralf Huss, and Günter Schmidt
Copyright © 2018 Pan Stanford Publishing Pte. Ltd.
ISBN 978-981-4774-88-8 (Hardcover), 978-1-351-13427-9 (eBook)
www.panstanford.com

imaging-based artifacts while including pertinent regions of interest to answer clinical questions. As these algorithms' complexity advances, they will not only free up pathologists' time but also unravel clinically relevant, yet undetected, complex morphological features and cellular interactions. Big data is currently employed in many aspects of medicine, and digital pathology will be no different. The wealth of data that image analysis can now produce, coupled with clinical and molecular data, will forge the path toward a personalized prognostic and predictive pathology.

9.1   Introduction

The field of digital pathology is currently experiencing one of the most exciting periods since its conception. It is rapidly transitioning from primarily a research tool into a viable clinical solution for patient primary diagnosis and determination of prognosis. This is a huge step-change from the microscopic evaluation and archiving of tissue sections, mounted on glass slides, which has been routine in the clinic since the 19th century. There are multiple reasons for this current clinical adoption of the technology, including technical advances in information technology infrastructure and whole-slide scanners, validation of the technology, regulatory approval, and the willingness of pathologists themselves.

As the field is still in its infancy, there are multiple terms for the different aspects of the technology. In this chapter, we refer to digital pathology as encompassing the entire infrastructure needed to effectively view and manage digital images of tissue sections within a clinical setting; from scanners through to laboratory information systems and digital archiving of specimens. While whole-slide imaging is the high-resolution digitization of whole tissue sections mounted on glass microscopy slides and is arguably one of the most essential components for the success of digital pathology, without high-quality images, the field falls at the first hurdle. These digital images can be viewed and navigated using ergonomic high-definition computer interfaces while being easily shared with colleagues on a global scale with the click of a button.

Digital pathology is, in essence, a disruptive technology that requires a complete overhaul of the workflow of traditional histopathology departments. This includes costly acquisition of the required technology and a change in mindset and working practice for the histopathologist. Unlike radiology, the transition to the digitization of clinical samples has been a slow one and is still very much ongoing. The vast majority of pathologists continue to view patient specimens using traditional microscopy; however, as more regulatory approval follows on from validation studies, this is beginning to change. The early adopters of digital pathology have been in education, research, and the pharmaceutical industry, where its advantage of remote viewing and sharing of slides coupled with the ability for automated analysis has pushed the technology forward. For the technology to be fully implemented, it must prove to be standardized, cost-effective, and improve upon current practice in both workload and time efficiency, such as improved ergonomics, easier navigation of the slide, remote reporting, and inclusion of dynamic slide images embedded within the digital pathology report. Countries where there may be large distances between specialist pathologists were the first to pass regulatory clearance for primary diagnosis, for example, in Tetu and Evans (2013) and Thorstenson *et al.* (2014).

Digital pathology proves advantageous not only in the morphological observation of histopathology specimens but also in the field of molecular pathology. Scanner vendors are incorporating multiplexed immunofluorescence and bright-field capability to allow dual functionality in a single scanner. Although still in the realm of research, the future of digital pathology will incorporate multiplexed immunofluorescence and automated image analysis. Multiplexed immunofluorescence enables the pathologist to view multiple molecular markers co-localized to a single cell while incorporating the cell's morphology. Automated image analysis of the digital tissue section will initially take the form of simple quantification of biomarkers but will move forward to include capturing morphological measurements and employ deep machine learning.

9.2 Brief History of Digital Pathology

The origins of digital pathology lie in the field of telepathology, whose own origins can be dated back to the 1960s but took hold in earnest in the 1980s. Telepathology refers to the practice of pathology at a distance, where the pathologist is remotely located from the actual specimen. This practice was initially used for educational purposes, for second opinions, or expert review when a specialist may be located some distance from the primary point of care (Farahani and Pantanowitz, 2015). Initially telepathology relied on sending static images from camera-mounted microscopes to colleagues in remote locations. This changed in the 1990s when dynamic imaging was possible with robotic driven microscopes, which could be controlled remotely. Dunn *et al.* showed that remote diagnosis using their robotic system saved both time to diagnose a patient and held a high correlation with traditional microscopic diagnosis (Dunn *et al.*, 2009).

From the late 1990s and onward, whole-slide scanners have been the tool of choice in the field of digital pathology and have enabled the technology to move the field from being based on remote access to a microscope to the preparation and interpretation of digitized tissue images for primary diagnosis. Initial whole-slide scanners were too slow and created too poor a quality of image to use routinely in the clinic. Recent technological advances have produced affordable scanners, which allow a high-resolution bright-field tissue section to be scanned in its entirety within a few minutes. Further technological advances incorporated into scanners allow bright-field imaging and multiplexed immunofluorescence capture within the same platform. Although the technology was available, regulatory hurdles had to be overcome in some countries, which changed as recently as April 2017 when Philips' ultra-fast scanner was passed by the US Food and Drug Administration (FDA) to be used for primary diagnosis (Caccomo, 2017).

9.3 Digital Pathology in Current Practice

9.3.1 Research

In the post-genomic era, the importance of spatial localization of gene expression and epigenetic heterogeneity is increasingly recognized as a missing link, bringing order and sense to complex, multiscale datasets based on an assessment of the genome. Together with advances in microscopy, there has been a huge challenge for cell biologists, initially to simply curate large numbers of images and allow easy retrieval and comparison with other data sources (Allan *et al.*, 2012). This aspect of digital imaging has probably advanced faster than any clinic-based strategy, but it is limited to essentially forming a repository of easily accessible, catalogued images. In the pharmaceutical industry, for example, with toxicological screening, several geographically distinct sites might be involved in a single study. Digitization of tissue images, curation, and rapid access have allowed international corporations to more effectively share data, facilitate discussion, and, therefore, conserve and strengthen expertise. The application of machine learning or automated image analysis to such images has been less rapidly adopted. This has, in part, been due to cultural acceptability, but more significantly the need to recognize and quantify abnormal features (*e.g.*, preneoplastic liver nodules in toxicological screening) rather than quantifying measureable, predefined features. The addition of immunohistochemical or immunofluorescence brings an additional layer of complexity but further increases the amount of information that can be extracted from tissue in the research environment, which will eventually translate into the clinic.

9.3.2   Education

Generations of biology and medical undergraduates have struggled with classroom microscopes to make sense of histology, eventually with the help of books and instructors, to accumulate sufficient ability to recognize patterns so that examinations become less of a challenge. What has been apparent that while some students will take to the subject matter, many are left wondering how pattern recognition can ever be regarded as a core discipline within modern omics-based science. One of the greatest benefits of the digitization of images has been the ability to annotate images, point out key features, and crucially explain the image in terms of the underlying dynamical processes that produce the observed visual features. This does not negate the value of students learning microscopy, particularly at a higher level, but it does mean that there is a quality assured, consistent, and truly integrated way of being able to introduce basic concepts. Furthermore, time in the microscopy lab and, in turn, time for students to familiarize themselves with histopathology, is limited. Digital pathology enables online tutorials, lectures, and unlimited access to teaching slides, thus greatly increasing a student's time for hands-on learning.

Clinical histopathology, to some extent, still relies on pattern recognition, but increasingly the diagnosis is supported by other investigations such as immunohistochemistry (IHC) and molecular genetics. In order that all of these strands can be integrated, there is increasing use of digital imaging in postgraduate training, both to instruct but also to ensure inter-center quality control and collegiality, making use of social media and Internet conferencing to foster a spirit of learning together and pushing the boundaries of what information is available in tissue. As the aims of pathology increasingly move from simply diagnosis and prognosis to prediction, the adoption of digital imaging, with the additional resource of image analysis, is gathering pace as part of a comprehensive approach to diagnostics in the broadest sense.

9.3.3 Clinical

As mentioned above, enthusiasts have used telepathology, effectively driving a microscope remotely, for several decades. There is now a revolution gaining pace to replace the much traditional microscopy with an interpretation of digital images. Whole-slide scanners are vital to modern digital pathology but are only one aspect of the whole workflow and information technology infrastructure. For a pathology lab to go fully digital, there must be a complete change of practice within the department. The digital laboratory information system allows full tracking, through barcoding, of each specimen throughout the entire workflow. This alone adds value to a digital system by automatically assigning a specimen to a specialist and by minimizing patient identification errors from accession through to the publication of the pathology report. Upon digitization, the slides can be automatically added to the department or individual pathologist's case load and viewed alongside digitized patient notes at individual workstations. This process minimizes the risk of misplaced slides, breakage of slides, and case notes and slides being misaligned.

The adoption of digital pathology is not simply as a response to the digital era, though the lessons learned from radiology's conversion to digital are not lost, but rather because the patient's histopathological assessment is intimately embedded in an electronic record that is not static. The application of systems biology thinking to pathology necessitates an ongoing, dynamic access to tissue, and all the data that are released from it, to best model the treatment for an individual patient. Largely led by industry, FDA approval for digital reporting has been granted, and a point of inflection has been reached where the question is no longer "should we adopt digital pathology?" but "when?" The next phase will be the adoption of comprehensive image analysis tools that refine and facilitate the process of quantifying features that influence prognosis or prediction. One envisages, rather like experience in genomics, that initially bespoke single solutions for particular needs will be developed and sold, to be followed by a more comprehensive, overarching *next-generation pathology* where data extraction and interpretation will parallel pattern recognition and candidate-led approaches to feature identification (Caie and Harrison, 2016).

9.4 Future of Digital Pathology

9.4.1 Mobile Scanning

Current digital pathology still relies on expensive infrastructure, not least the scanner itself. Many institutes where the advantages of digital pathology and remote diagnosis would benefit the most also suffer from a lack of funding, which makes such large-scale invest-ment impossible. As technology advances, it tends to become more compact as well as cheaper. Both the mobile phone and the satellite drone sectors are pushing compact, high-quality imaging techno-logical advances forward. This will reap benefits in digital pathology where devices are being developed that can be portable and rela-tively cheap compared to large-scale scanners. Such technology will allow patients to access global expert pathologists where no such local person exists. Technology exists that can use a mobile phone as a digital scanner, creating a low image file size making the shar-ing and storing amenable to low-resource countries (Auguste and Palsana, 2015). Another advance that relies on mobile phone technology is hand held and makes digital scanning truly portable. This device named *MoMic* uploads images through wireless technology into cloud servers and would allow hospitals and point-of-care sites to scan patient samples and send them to experts anywhere in the world (Lundin, 2016). Furthermore, patient samples could be digitized and sent to pathologists from within the surgical theatre, for example, to check surgical margins for the presence of cancer cells.

9.4.2 Feature-Based Image Analysis

Digital histopathology is still reliant on manual semi-quantitative and categorical reporting of complex tissue as dictated by institutional guidelines. Although pathologists are highly skilled at image processing and identifying significant features within the tissue sample, observation by eye is subjective and open to observer variability. The manual counting of features, such as tumor buds and immune infiltrate in colorectal cancer, or IHC results, such as KI67 or HER2 positivity in breast cancer, can be time consuming and is not highly reproducible. Furthermore, pathologists tend to only report on a subset of the tissue to save time; however, this subset is subjectively chosen. Increasingly, there is an awareness that reliable quantification of many features previously graded by an expert brings advantages of reproducibility, refinement of hypothesis testing (because variables can be more appropriately described as a continuum rather than just categorical) and incorporation of data previously unused because of the difficulty of an expert in describing it (*e.g.*, the metadata implicit in measurement of features that are stochastic or markedly heterogeneous).

Automated image analysis can overcome many restrictions associated with traditional scoring by a pathologist. These include negating observer variability by the standardized quantification of features across large patient cohorts, the reporting of continuous data across a whole tissue section at single cell resolution and identifying obscure or rare events. In this manner, image analysis can begin to report on the heterogeneity within a tissue sample and does not simply categorize an entire patient sample as either positive or negative.

The future of digital pathology will undoubtedly include automated image analysis. Many whole-slide scanners now include image analysis software as part of the digital pathology package.

The available software can vary from simplistic counting algorithms to the more complex, which can co-register features across serial tissue sections. Histopathology has always relied on assessing the morphology of cells and tissue but traditionally report these complex and often subtle cellular differences in a semi-quantitative and categorical manner. Image analysis and the evolving field of tissue phenomics not only captures precise morphological characteristics of each cell in a region of interest using continuous data, but also co-registers morphology to cell type and biomarker expression. Furthermore, multiple objects and cellular phenotypes can be analyzed and reported on depending on their spatial interaction within the tissue microenvironment (Fig. 9.1).

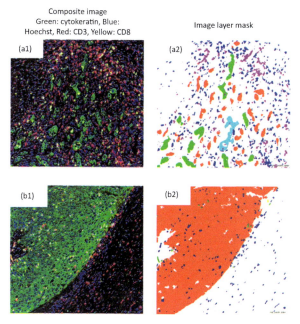

Figure 9.1 Examples of heterogeneous histopathological features in bladder cancer visualized by multiplexed immunofluorescence and analyzed by automated image analysis: (a1) Infiltrative invasive pattern and (b1) pushing border pattern of bladder cancer cells (green) co-registered with lymphocytic infiltration (red and yellow). (a2 and b2) Automated image analysis using Definiens Developer XD software to quantify subpopulations of tumor, the surrounding lymphocytic infiltration and their spatial relationship to each other. (a2) Red, tumor buds; green, large tumor buds; light blue, tumor mass; dark blue, CD3 cells; pink, CD8 cells. (b2) Red, tumor mass; blue, CD3 cells.

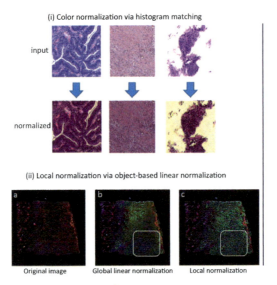

Figure 9.2 Image normalization allows accurate standardized automated image analysis of heterogeneous cohorts: (i) color normalization of H&E slides using a histogram matching algorithm. (ii) Automatic fluorescence normalization of a carcinoma with low and heterogeneously expressed cytokeratin (green). (a) Original image with cytokeratin intensity too low for accurate automated image analysis. (b) Global linear normalization results in heterogeneous intensity (white box). (c) Local reference objects are used with linear normalization to ensure a standardized and even intensity across the tumor.

For image analysis to be part of the pathologist's toolkit, there are certain obstacles, which must be recognized and addressed, principal among which is standardization. The human eye is excellent at accounting for variances in the imperfect tissue sample, but the computer is less so. Such variances and artifacts may stem from surgical ischemia, tissue fixation, section cutting, and labeling. Modern scanners are overcoming imaging artifact and attempt standardized whole-slide capture, although evenly illuminated and out-of-focus fields persist when batch imaging hundreds of samples in a single run. Many software vendors do not consider this and apply algorithms based on a set of thresholds alone, which indiscriminately count all objects within them, some of which could stem from artifact. The field of image analysis is beginning to recognize and address this issue. The more sophisticated image analysis software can now exclude autofluorescence, tears, and edge effect (Caie *et al.*, 2016), and account for inherent heterogeneity within patient samples. Non-standardized labeling of tissue and patient heterogeneity can affect the intensity and color of the resultant digital image. Algorithms can now automatically adjust for these variances and create standardized images across all patient samples. These standardized images are far more amenable to allowing accurate quantification of batch-processed samples with a single fully automated algorithm (Fig. 9.2).

9.4.3 Machine Learning on Digital Images

The brain is excellent at processing complex images and pattern recognition while negating tissue artifact and non-standardized color intensities, all essential factors to accurately report on patient samples. Traditional threshold-based automated image analysis may struggle with these artifacts and report false positives. However, the ability of the most recent machine learning algorithms to *learn* from complex digital pathology images by including texture and context of pixel-based objects has enabled the automatic exclusion of artifact, such as necrosis, and the inclusion of tissue containing features of interest, such as the invasive margin of a tumor (Fig. 9.3).

Deep machine learning on digital images can not only automatically classify regions of interest, such as stromal or tumor segmentation, but can also identify finer detail within the tissue such as cell types (Brieu *et al.*, 2016), their expression profiles and their neighboring local environment, all of which may contain significant information to answer specific clinical questions. In fact, the deep learning algorithms are reaching an accuracy that is at a level comparable to human perception and will undoubtedly be able to report on subtle but pertinent morphological or spatial relationships of cells that are hitherto undetectable or reproducible when quantified by the human eye. There are different approaches to machine learning, which are addressed in Chapter 4, but all of them come with caveats. Machine learning must be performed on a sufficient number of training samples from a wide source to avoid overfitting and for it to remain accurate across large *unseen* patient populations (Cruz-Roa *et al.*, 2017). The annotated training samples must be free from selection bias and also representative of a wider population. Finally, the ground truth used for training the algorithm must be accurately classified and as this is usually performed by a human carries a certain inherent error rate. The ultimate goal of machine learning within the future of digital pathology is to be able to apply a single algorithm across inter- and intra-heterogeneous patient samples sourced from institutions across the globe while reporting accurate diagnostic, prognostic, or predictive results.

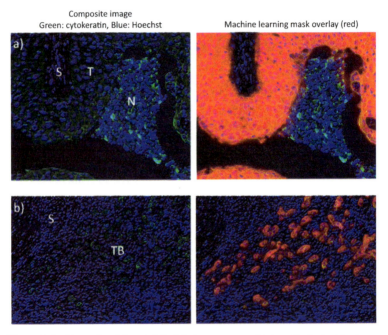

Figure 9.3 Examples (a) and (b) of heterogeneous histopathological features in bladder cancer visualized by immunofluorescence and analyzed by machine learning. Machine learning is trained to identify features of interest such as tumor bulk (T) and tumor buds (TB) and to disregard the stroma (S) and necrosis (N).

9.4.4 Big Data and Personalized Pathology

Pathology has embraced the molecular era and is producing big data across multiple omics platforms. The majority of these do not take into account tissue architecture. In pathology, context is everything. Advances in digital pathology and tissue phenomics now allow the generation of big data across intact tissue sections. Image analysis already allows the co-registration of cellular shape and extent with subcellular localizations of biomarkers, while placing objects in spatial context. Advances in the labeling and scanning of multiplexed immunofluorescence can add to the number of data-points one can capture. Furthermore, tissue co-registration software allows the co-localization of many more aspects of the tissue microenvironment, not just of the diseased cells but also of the host reaction in response to the disease. The big data from tissue phenomics is a perfect match to be incorporated with other omics datasets. Machine learning on the big data itself can utilize all available parameters, or be employed to identify the optimal parameters and associated clinical cutoffs to answer a specific clinical question (Caie *et al.*, 2016; Yu *et al.*, 2016). In this manner, thousands of data-points may be distilled down to a handful of pertinent parameters. These few parameters are far more amenable to a clinically applicable test. Therefore, the future of pathology may rely on complicated research-based big data, which can be simplified down to translate into the clinic. The use of these personalized big-data sets will lead the way for the future treatment of patients in a personalized and precise manner.

References

Allan, C., Burel, J. M., Moore, J., Blackburn, C., Linkert, M., Loynton, S., Macdonald, D., Moore, W. J., Neves, C., Patterson, A., *et al.* (2012). OMERO: Flexible, model driven data management for experimental biology. *Nat. Methods*, **9**, 245–253.

Auguste, L. and Palsana, D. (2015). Mobile whole slide imaging (mWSI): A low resource acquisition and transport technique for microscopic pathological specimens. *BMJ Innov.* Published Online First: 10 June 2015. doi: 10.1136/bmjinnov-2015-000040.

Brieu, N., Pauly, O., Zimmermann, J., Binnig, G., and Schmidt, G. (2016). Slide-specific models for segmentation of differently stained digital histopathology whole slide images. In *Proc. SPIE 9784, Medical Imaging 2016: Image Processing, 978410* (21 March 2016); doi: 10.1117/12.2208620.

Caccomo, S. (2017). FDA allows marketing of first whole slide imaging system for digital pathology. FDA News Release. www.fda.gov/newsevents/newsroom/pressannouncements/ucm552742.htm

Caie, P. D. and Harrison, D. J. (2016). Next-generation pathology. *Methods Mol. Biol.*, **1386**, 61–72.

Caie, P. D., Zhou, Y., Turnbull, A. K., Oniscu, A., and Harrison, D. J. (2016). Novel histopathologic feature identified through image analysis augments stage II colorectal cancer clinical reporting. *Oncotarget*, **7**, 44381–44394.

Cruz-Roa, A., Gilmore, H., Basavanhally, A., Feldman, M., Ganesan, S., Shih, N. N. C., Tomaszewski, J., Gonzalez, F. A., and Madabhushi, A. (2017). Accurate and reproducible invasive breast cancer detection in whole-slide images: A deep learning approach for quantifying tumor extent. *Sci. Rep.*, **7**, 46450.

Dunn, B. E., Choi, H., Recla, D. L., Kerr, S. E., and Wagenman, B. L. (2009). Robotic surgical telepathology between the Iron Mountain and Milwaukee Department of Veterans Affairs Medical Centers: A 12-year experience. *Hum. Pathol.*, **40**, 1092–1099.

Farahani, N. and Pantanowitz, L. (2015). Overview of telepathology. *Surg. Pathol. Clin.*, **8**, 223–231.

Lundin, J. (2016). Mobile digital microscopes are in development. *Eur. Hosp.*, **26**, 1.

Tetu, B. and Evans, A. (2013). Canadian licensure for the use of digital pathology for routine diagnoses. *Arch. Pathol. Lab Med.*, **138**, 302–304

Thorstenson, S., Molin, J., and Lundström, C. (2014). Implementation of large-scale routine diagnostics using whole slide imaging in Sweden: Digital pathology experiences 2006–2013. *J. Pathol. Inform.*, **5**, 14.

Yu, K. H., Zhang, C., Berry, G. J., Altman, R. B., Re, C., Rubin, D. L., and Snyder, M. (2016). Predicting non-small cell lung cancer prognosis by fully automated microscopic pathology image features. *Nat. Commun.*, **7**, 12474.

Chapter 10

Tissue Phenomics in Clinical Development and Clinical Decision Support

Florian Leiß and Thomas Heydler

Definiens AG, Bernhard-Wicki-Strasse 5, 80636 Munich, Germany
fleiss@definiens.com

10.1 Cancer and Oncology Drug Development

Cancer affects a large and growing number of patients every year. Ageing populations in developed countries and improving overall health in developing countries result in rising numbers of newly diagnosed cases. More and more people live with cancer as tumors are detected earlier and therapies have improved for some cancers. Progress is very limited for other types of cancer with very poor prognosis and few promising treatment options. Cancer-related deaths are on the rise around the world.

The development of oncology drugs is a focus of the pharmaceutical industry. Hundreds of novel therapies are in clinical development. Development costs have exceeded one billion dollars per newly launched medicine and keep rising. The need to recover these

Tissue Phenomics: Profiling Cancer Patients for Treatment Decisions
Edited by Gerd Binnig, Ralf Huss, and Günter Schmidt
Copyright © 2018 Pan Stanford Publishing Pte. Ltd.
ISBN 978-981-4774-88-8 (Hardcover), 978-1-351-13427-9 (eBook)
www.panstanford.com

investments and maintain or re-gain profitability puts economic pressure on the pharmaceutical industry. Consequently, oncology drug prices increase and overall spending on cancer care surge. Outcome-based reimbursement could be a solution but the required approaches and mechanisms remain to be established.

Two main trends will significantly change our approach to deal with cancer: immunotherapy and the digitization of healthcare. They individually raise hopes for major improvements but pose new challenges of their own. They complement each other and can jointly help to overcome these new challenges to establish the future of cancer care.

10.2 Immunotherapy

Immunotherapy has changed the way we treat cancer and how we think about cancer. With checkpoint inhibitors (anti-PD-L1, anti-PD-1, anti-CTLA-4) at the forefront, new classes of drugs have entered center stage in cancer therapy. Remarkable successes in clinical trials and lasting benefits for some patients raise hopes for a breakthrough in cancer entities that did not see much progress in decades. Conceptually, immunotherapy has shifted our focus from the tumor itself to the immune response against it. In some cases, removing or killing cancer cells directly appears less effective than enabling the patient's immune system to attack or contain tumor cells.

Deepening our understanding of a patient's immune status has become a prime objective in medical science. It remains poorly understood why some patients respond to the novel therapies while others do not. It will be key to identify the limitations of the immune response to provide the right help to the patient. Tissue is known to be an important source of the required information. Types, densities, and spatial relationships of immune cell populations (*e.g.*, cytotoxic T cells, regulatory T cells, macrophages) in the tumor microenvironment (TME) are known to have prognostic and predictive value (see Chapter 6). Expression of PD-L1 is predictive of response to checkpoint inhibitors (*e.g.*, anti-PD-1 or anti-PD-L1), and assays to assess PD-L1 expression status are now used in clinical routine. Unfortunately, these assays are not perfect. Some test-negative patients do respond, and many test-positive patients do not.

Existing assays certainly provide limited guidance to determine the optimal treatment path as novel agents will be approved. Early clinical evidence and our mechanistic understanding of immunotherapy agents suggest that combination therapy will be the future. Different approaches to enable the immune system can be combined (*e.g.*, checkpoint inhibition, vaccination, CAR T-cell therapy) with each other, with existing treatment modalities (surgery, radiation, chemotherapy), and other targeted drugs (*e.g.*, anti-EGFR, anti-KRAS, anti-ALK, anti-BRAF). While very promising, combination therapy raises novel challenges for our approaches to evidence-based medicine and personalized medicine.

Carefully conducted clinical trials (randomized, double-blind, multi-centric) are at the beacons of evidence-based medicine. They should guide treatment decisions. But as combination therapy dramatically increases the number of treatment options, there will not be enough clinical trials. There are too many options to combine the emerging therapies to test them all. Physicians and patients will increasingly face decisions that are not backed up with the gold standard of evidence-based medicine. Pharmaceutical companies can demonstrate safety and efficacy of novel drugs. But it is beyond their financial capabilities (and interests) to compare all available treatment options.

The prevailing approach to personalized medicine adds to this challenge. Bringing the right therapy to the right patient at the right time all too often starts with the therapy. Diagnostics are developed to identify patients that are most likely to respond to a novel drug (companion diagnostics or complementary diagnostics). Clinical trials demonstrate a favorable risk–benefit ratio for patients selected with the new diagnostic. This makes it more difficult to compare two novel treatments. Wherever different diagnostic tests have been used to select patients, differences between patient populations are very likely. These differences make it hard to conclude which of the two treatment options is superior.

Personalized medicine should start with the patient. Diagnostics should empower physicians to determine the best course of treatment. Companion diagnostics are not designed to do that. They can establish eligibility for one course of treatment (one drug, one test). Diagnostics for the future of personalized medicine should provide information that is relevant across treatment options (one test, many drugs). The medicine of the future should be evidence-based. The evidence should enable comparisons between treatment options and enable rational decisions for combination therapy. The digitalization of healthcare represents an opportunity in both regards.

10.3   Digitalization of Healthcare

Digitalization is transforming healthcare. Barriers to change slowed the adoption of information technologies in healthcare as compared to other industries. These include the complexity of interrelations between many different stakeholders and strong regulation to protect patient safety. With some delay, some of the familiar patterns of digitalization are now emerging in healthcare.

The abilities to collect and analyze data are the gatekeepers of the new era. The amount of data being generated in healthcare is growing rapidly. New devices capture new types of data (*e.g.*, whole genome sequences) and old types of data more often (*e.g.*, continuous tracking of heartbeat with wearables). Costs for data storage and computation are decreasing rapidly. Data analytics is advancing and scientific progress accelerating.

The explosion of information results in a complexity crisis in healthcare. Physicians cannot possibly integrate all available information on a patient and relate it to the body of medical knowledge in the way they did in the past. Integrating available information is challenging as additional sources of information emerge, data formats vary, and the data volume increases. Relating information to medical knowledge is challenging as new insights are rapidly emerging. Knowledge has become very dynamic. Even specialists are struggling to stay up-to-date with increasing numbers of scientific publications and clinical studies.

Information technologies are at present more successful in creating the complexity crisis than at solving it. Our abilities to collect health-related data are currently outpacing our abilities to analyze them. Yet they will eventually need to support physicians in making sense of the data and in establishing actionable recommendations. Analytics are at the core of other successful digital transformations.

Information technologies will eventually help to solve the complexity crisis in healthcare that they helped to create. By improving the profiling of patients and identifying the best matches among available treatment options, they will put the patients (not the drugs) at the center of personalized medicine.

10.4 Patient Profiling

Collecting information about a patient's tumor and the immune response against it should be the first step in personalized cancer care. The information will include biomarkers with known clinical relevance and should be broad enough to be relevant across treatment options (one test, many drugs). Genetic information on the tumor (mutations, rearrangements, translocations, mutational load) has known prognostic and predictive values (*e.g.*, for targeted therapies against EGFR, KRAS, ALK, ROS1, or BRAF alterations). Protein biomarkers (*e.g.*, HER2, PD-L1) are relevant for treatment decisions. Densities and locations of immune cell populations in the TME (*e.g.*, cytotoxic T cells, regulatory T cells, macrophages) are also known to be associated with outcomes and therapy response (see Chapter 6).

Patient profiles must be established as consistently as possible to enable systematic comparisons between differing profiles (prognostic information) and different treatment options for patients with similar profiles (predictive information). They should be comprehensive enough to enable the discovery of biomarkers that are not known initially. Progress with genomic sequencing enables consistent genomic profiling at rapidly decreasing costs. The resulting data is digital and can be included in standardized patient profiles. Additional genomic biomarkers beyond those of known clinical value can be included to enable the discovery of novel prognostic and predictive markers.

The consistent profiling of a patient's immune response is possible but less well established than genomic profiling. Biomarkers with known and suspected clinical value can be visualized within the context of intact tissue using immunohistochemistry or *in situ* hybridization. A current limitation for systematic profiling in clinical routine is the visual assessment of tissue biomarkers.

Pathologists are very skilled at detecting and interpreting clues in tissue morphology and protein expression but less so at providing consistent quantification. Computer-based systems can distinguish, localize, count, and spatially relate thousands of immune cells in a digitized tissue section to extract quantitative information that is impossible to capture manually by pathologists (see Chapters 2 and 3). A standardized set of tissue biomarkers with corresponding detection algorithms can be established to determine a fingerprint of a patient's immune response (immuno-oncology panel). Multiple tissue biomarkers with known and suspected clinical value can be included. Together with a set of genomic markers (and potentially other types of biomarkers in the future), the resulting information constitutes the patient profile.

Broad and consistent patient profiling can help to make different clinical trials more comparable. It could enable pharmaceutical companies to take better portfolio decisions by allowing data-driven selection of drug candidates and cancer indications. Trial and error can be reduced by supporting rational choices of drug combination programs. An established panel for patient profiling could facilitate stratification in early clinical trials to reduce the trial size and improve risk–benefit profiles. It can enable innovative and adaptive trial designs (*e.g.*, by combining characteristics of the umbrella and basked trials). A broad diagnostic workup including multiple genomic and phenomic biomarkers could help to improve the understanding of a drug's mechanism of action, its pharmacodynamics, tumor heterogeneity, and mechanisms of therapy evasion. Jointly these effects could contribute to a reduction in pharmaceutical R&D spending or increased output and help to bring novel drugs to market faster.

Beyond drug development, patient profiling could enable physicians to personalize treatments. Many treatment decisions involving combinations or sequences of novel treatments will not be guided by clinical trials that establish the selected choice as the best one. But profiling can help to collect the information required to learn from each treatment decision to provide a better basis for subsequent decisions. Again, it is key that patient profiling is consistent and sufficiently broad. Measuring different sets of biomarkers or the same biomarkers with too much inter-observer variability makes it difficult to compare profiles across larger cohorts of patients. A broad set of biomarkers will be needed to cover relevant information across treatment options and to identify new prognostic or predictive biomarkers.

10.5   Therapy Matching

Establishing comprehensive and consistent patient profiles provides the basis for a systematic evaluation of correlations between biomarkers and patient outcomes. Powerful analytic capabilities are required and have been developed to identify patterns in genetic information. Point mutations, deletions, translocations of oncogenes, and tumor-suppressor genes that are known to carry prognostic or predictive information are examples.

Similarly, quantitative fingerprints of a patient's immune responses need to be compared with each other and correlated with outcomes and genetic information. The visual examination of tissue samples by pathologists remains the gold standard in cancer diagnostics despite the rise of novel bioanalytical methods. Morphological characteristics of cancer cells and spatial distributions of immune cells can only be assessed from an intact tissue. However, pathologists cannot routinely provide the rich quantitative readouts that enable powerful analytics to reveal novel correlations. These limitations can now be overcome with computer-based systems that help to capture the information from tissue samples in digital formats for subsequent data processing. Consequently, patterns in interactions between different immune cell populations and tumor cells with strong correlations to outcome and therapy response will likely be discovered. These will complement existing (*e.g.*, genomic) and other novel (*e.g.*, proteomic) biomarkers to better guide therapy decisions in the future.

Large cohorts of patients are required to reveal a subtle patterns. Clinical trials are an important source of information, and comprehensive patient profiles can be established in clinical trials to better distinguish between responders and non-responders to the therapy under investigation. However, many relevant treatment options will not be tested in clinical trials as opportunities to combine treatments increase. A novel approach is needed to avoid a therapeutic anarchy where physicians revert to gut feelings and beliefs in the absence of sound data.

Clinical trials provide the best source of evidence but not the only source of evidence. They are superior to observational studies as they can rule out more potential confounders if appropriately designed. Expectations of patients and physicians are less likely to affect study results if randomization, double-blinding, and placebo controls are used (they can avoid patient selection bias, treatment bias, placebo effect, and accidental un-blinding). But there are limitations to a paradigm that relies on careful studies with a few patients to guide treatment decisions for many. As treatment decisions are becoming increasingly complex, we will need more data points to learn from. A simple increase in the numbers of clinical trials as currently seen in immuno-oncology will not be an adequate solution. First, results of different trials cannot be easily compared if different enrollment criteria are used, and the trend to use different companion diagnostics for stratification makes this problem more challenging. Second, well-designed clinical trials are very expensive. It is not sustainable for health systems to pay for ever-increasing numbers of clinical trials (mostly indirectly via high prices for novel treatments).

Dynamic algorithms to optimize the matching between individual profiles and available options will be the solution. We will need to learn from the effects of all treatment decisions, not only those in clinical trials. Some patients will not want to be part of a study that randomly assigns them to one of multiple treatment options (including placebo). We can thus only observe how patients respond to the treatments they and their physicians chose. But comprehensive patient profiling will allow comparing similar patients that receive different treatments. Observational studies can be powerful. Wherever the observed effect size rules out the combined effects of plausible confounders, they are highlighted as a sound source of evidence by proponents of evidence-based medicine.

Predictive analytics cannot determine optimal treatment decisions. Physicians and patients must be in control. And it certainly makes the most sense start with the best hypotheses we already have. Clinicians have a lot to teach emerging clinical decision support systems (DSSs). But algorithms can learn from leading experts to make medical knowledge accessible to others. They can list treatments that were successful at treating patients with similar profiles. They can query observational data to compare two treatment options in the absence of an adequate clinical trial that supports a decision. Eventually, novel hypotheses can be generated based on newly discovered correlations.

Individual patients are at the center of treatment decisions. As much information on the patient and medical knowledge as possible should be considered. Information technologies should enable physicians to deal with the complexity. Populations of patients are the focus of drug development. Analytics can help to identify groups of patients with an unmet medical need and similarities in their profiles. Systematic surveys of characteristic immune responses in different cancer indications could suggest which drug combinations are most promising in which indication. They can help reveal differences between responders and non-responders to better understand mechanisms of action and guide future drug development.

Ideally, clinical research and clinical development should be integrated with patient profiling in clinical routine. Patients with profiles that suggest limited benefits for established treatment options are prime candidates for enrollment into clinical trials. Patient profiles could be used to determine eligibility for enrollment into clinical trials. This would extend therapy matching for patients to experimental drugs in clinical development. At the same time, it could save time and money spent by pharmaceutical companies to recruit patients into their trials. Clinical research could be conducted on larger patient cohorts wherever patients consent to the use of their data to advance research.

10.6 Benefits

Patient profiling and therapy matching will help to create the future of cancer care. They will bring together the promises of immuno-therapy and digital healthcare. They will enable novel interactions between the key stakeholders in healthcare to realize network effects as seen with marketplaces, platforms, and ecosystems in other industries. Progress may be slowed by the complexities of healthcare and the interest of some players to defend the status quo. But eventually new approaches and business models will transform healthcare. All stakeholders will have strong benefits.

First and foremost, patients will receive better cancer care. Outcomes will improve if therapies can be better personalized to a patient's tumor and immune response. Patients will have access to more information relevant to their treatment decision. Their physicians' recommendations will be more transparent. They will be enabled to better understand their options and to take more ownership of treatment choices. Physicians will receive consistent and quantitative information on a patient's profile and references to successful treatments in similar cases. Approved treatment options can be ranked, eligible clinical studies can be listed, and relevant clinical evidence can be summarized. More physicians could be enabled to identify the best approved or experimental therapies for patients at a level currently restricted to leading academic centers.

Pharmaceutical companies could identify patient populations with benefits from new drugs earlier in development to shorten time to market and reduce R&D spending. Consistent biomarker profiles across clinical trials will support rational decision making in selecting drug combinations and most promising indications. Digital technologies can help to better engage with patients as their primary customers. Payers could obtain data to compare the effectiveness of different treatment options to enable outcome-based reimbursement. Health systems can learn to spend money more efficiently to contain rising costs for healthcare. Regulators could identify rare adverse events earlier to better protect patients at risk. Real-world evidence could mitigate some of the well-known risks of translating results from clinical studies with highly selected participants (*e.g.*, below a certain age, absence of typical co-morbidities) to a broader population.

10.7 Conclusion

Improving patient profiling and therapy matching will be key to escape the one drug–one-diagnostic-trap to truly personalize cancer care. They will enable us to deal with the challenges of rapidly evolving knowledge in immuno-oncology and emerging combination therapies. Patient-centric solutions that blend the boundaries between drugs, diagnostics, and information technologies are required to improve the lives of many millions of patients affected by cancer.

Glossary

Big Data Analytics

The interactive, real-time analysis and visualization of huge datasets from heterogeneous sources, such as tissue image analysis results, gene expression, and clinical trial data. The datasets may comprise many rows, for example, hundreds of millions of cells detected in a clinical study, or a large number of columns, for example, all features that may represent a patient. Fast spatial search, feature selection, and advanced visualizations are the core components of such a software system.

Biomarker

According to an NIH working group, a characteristic that is objectively measured and evaluated as an indicator of normal biological processes, pathogenic processes, or pharmacologic responses to a therapeutic intervention.

Cancer Immunotherapy

The treatment of cancer by inducing, enhancing, or modulating an anti-cancer immune response. The status of the immune system in tissue, such as the spatial relationship of the tumor and immune cells, has to be assessed to determine patients' eligibility.

Cognition Network Language (CNL)

A graphical, interactive scripting language providing a rich set of instructions to implement context-driven image analysis algorithms following the principles of Cognition Network Technology.

Cognition Network Technology (CNT)

A knowledge-driven image analysis approach, grouping pixels in a digital image to meaningful objects and organizing these objects in a hierarchical semantic network, allowing thereby a context-sensitive analysis of complex biological structures such as the tumor microenvironment.

Computer-Aided Diagnosis

Software solutions that leverage the strength of informatics and computation to assist pathologists and physicians to make better diagnostic decisions with higher confidence.

Co-registration

The process of virtually aligning spatially related images to make them computationally available in one common coordinate system. Images may originate from differently stained consecutive sections of a tissue block or from any other imaging modality.

Cross-validation

Technique for model validation to estimate the generalization performance of a computational model for an independent dataset. The core of the method is the use of disjoint data subsets for training and testing of the model to limit overfitting. Data subsets are generated, for example, by systematic splitting into k subsets (k-fold cross-validation), by repeated random sub-sampling (Monte Carlo cross-validation), or combinations of both.

Data integration

In the field of bioinformatics, data integration refers to combining research results from multiple sources and linking them to clinical data, including disease progression data, using unique identifiers for samples and patients.

Decision Support System (DSS)

Methods, tools, or technologies (including software) that follow principles, theories, and concepts to enhance and improve decision making.

Digital Pathology

An information technology environment based on the digitization of tissue sections and supportive data to enable virtual microscopy and image analysis. It is the basis to improve and accelerate biomarker discovery, diagnosis, and prognosis in diseases such as cancer.

Feature Selection

The process of phene discovery: the mathematical process to select the most predictive measurement (feature) that represents the medical condition of a patient at a time point prior to a clinical decision. The prediction quality is characterized by predictive values, prevalence, Kaplan–Meier log-rank test, or other statistical evaluations on the cohort of patients in the study.

Image Analysis

The application of computational methods to extract meaningful information from digital images, including image processing, model-based, and machine learning approaches.

Image Mining

The process of generating knowledge from information implicitly available in images within exploratory data analysis.

Immunohistochemistry (IHC)

Chromogenic or fluorescent visualization of antigens on tissue sections through the selective binding of specific antibodies.

Invasive Margin (IM)

The outermost region of neoplastic cells in a tissue section, beyond which no other tumor cells are seen histologically or likely to exist.

Machine Learning

Algorithms that learn from data and make data-driven predictions by building computational models based on input data and expert knowledge, closely related to pattern recognition and computational learning theory, computational statistics as well as mathematical optimization. Machine learning includes supervised learning, that is, learning from examples with known labels, and unsupervised learning, that is, finding patterns in data without known labels. Popular examples of machine learning approaches are deep learning, for example, convolutional neural networks, or decision tree learning, for example, random forests.

Medicine 4.0

Includes the full digitization of all patient-related healthcare information, that is diagnostic image data, monitoring data from wearable devices, and patient medical records, enables big data analytics to improve therapy decisions.

Omics

A collective term that typically refers to genomics, transcriptomics, proteomics, metabolomics, tissue phenomics, and radiomics studies.

Pathology 4.0

The additional use and implementation of digital solutions such as tissue phenomics into tissue diagnostics and the routine pathology workflow.

Tissue Biomarker

A single indicator or a combination of markers to be informative on biological or pathogenic processes on or from tissues in a quantitative and/or qualitative manner.

Tissue Heatmap

An artificial, computer-generated low-resolution representation of a (virtually multiplexed) tissue section, in which each pixel represents some statistics of detected cells in the corresponding region in the high-resolution image.

Tissue Phene

A tissue phene is a quantitative descriptor (diagnostic algorithm) for functional, morphological, and spatial patterns in cells, tissues, and organs of biomedical relevance. Phenes may be composed of other phenes and include scores from histopathology, as well as clinical or omics variables.

Tissue Phenomics

Tissue phenomics is the systematic discovery of phenes, which correlate with disease progression. Resulting signatures find application in patient profiling and therapy decisions.

Tumor-Infiltrating Lymphocytes (TILs)

Immune cells, mainly T cells, which left the circulation and migrate into the tumor, where they are involved in the destruction of cancer cells. Their presence can be associated with a favorable clinical outcome.

Tumor Microenvironment (TME)

The spatially adjacent environment of a tumor, which is composed of fibroblasts, blood vessels, immune cells, and extracellular molecules.

It can experience a strong influence by the tumor and exert a strong influence toward the tumor, affecting immune responses on one side and tumor evolution on the other side. The tumor microenvironment is a key factor in cancer progression.

Whole-Slide Imaging

The procedures and technology to create and process high-resolution scans of histopathological glass slides.

Index

accuracy 76, 105, 108, 112, 116, 178, 180, 181, 195
activation function 87, 88, 92
aggregated information 106
agreement 26–30, 32, 33
algebraic feature composition 111–113, 120, 121
algorithm 37, 42, 54, 56, 87, 89, 91–93, 104, 105, 112, 140, 194–196
algorithmic framework 20
annotated training data 23
annotation 15, 16, 24, 27, 29, 42, 105, 126, 128, 133
anti-PD-L1 105, 151, 159, 200
a priori knowledge 24, 71
assay conditions 14
autofluorescence 13, 194
automated analysis 32, 67, 187
automated classification assessment 27
automated image analysis 109, 111, 159, 181, 185, 187, 189, 192–195
automatic quantification 5
autostainer 14

background staining 14
batch processing 24
big data 3, 4, 7, 10, 11, 126, 128, 162, 167, 176, 181, 182, 186, 196, 197, 209, 212
big data analytics 7, 209, 212
bioinformatics 7, 129–134, 136, 138, 140, 142, 144, 146, 148, 150, 152, 154
biological pathways 132, 137

biological questions 12, 21, 25, 136
biomarker 11, 12, 14, 147, 148, 151, 152, 154, 163, 165, 167–169, 171, 179, 181, 182, 208, 209, 211
black swan 24
Bland–Altman plots 29, 30
breakage of slides 190
breast cancer 32, 99, 100, 148–155, 163–167, 172, 173, 179, 180

cancer
 genesis 150
 immunology 5
 invasion 21, 33
 morphology 166, 170
 progression 119, 214
CD3 32, 108, 111, 112, 142, 160, 163, 164, 168, 169, 193
CD4 32, 155, 166, 168, 173
CD8 32, 33, 104, 105, 109, 112, 142, 153, 159, 160, 163, 166, 168–170, 173, 193
CD20 166, 168, 169
CD38 166
CD45RO 163, 168
CD163 160, 161, 164, 168
cell
 annotations 23, 27
 clusters 139
 contact 139
 densities 21, 65, 105, 160, 163, 170
 interactions 6, 50, 102, 129, 131, 139, 140

population 6, 22, 101, 114, 137–140, 158, 163, 200, 203, 205

receptor status 150

cell-to-cell distance 138, 139

checkpoint inhibitors 151, 200

CK18 160, 161

class assignment 20

class hierarchy 42–44

classification 17–19, 27, 29, 46–49, 70, 72–74, 77, 78, 82–84, 94, 95

client–server architecture 24

clinical adoption 186

clinical data 103, 113, 122, 123, 162, 177, 210

clinical history 176, 177

clinical information 132, 137, 143, 144, 150

clinical samples 187

clinical setting 186

clinical solution 186

clinical staging 162

clinical translation 185

clinical trial 114, 116, 127, 147, 163, 174, 207, 209

clonal heterogeneity 131

cloud environment 181

clustering 83, 111, 117–119, 129, 144

CNL *see* Cognition Network Language

CNT *see* Cognition Network Technology

Cognition Network Language (CNL) 18, 35–45, 48–50, 65, 66, 209

Cognition Network Technology (CNT) 7, 14, 16, 17, 21, 22, 31, 32, 67, 158, 159, 164, 168, 209, 210

cognitive digitization 6

cohorts 135, 137, 146, 150, 167, 192, 194, 205, 207

cold ischemia 14

colorectal cancer 142, 160, 162, 163, 168, 171, 172, 179, 192, 198

combination therapies 10, 159, 169, 208

combined biomarkers 143

common coordinate system 109, 210

companion diagnostics 8, 15, 115, 201, 206

comparability 27, 30, 31

complementary diagnostics 147, 201

complex tissue 185, 192

computed tomography 104

concordance 30, 33

confidence 30, 118, 140, 145–147, 183, 210

consensus 25, 27

context 10, 11, 35–40, 42–46, 49, 50, 54–58, 64–66, 84, 85

context-derived patterns 20

context-driven analysis 35, 37–40, 42, 43, 45, 46, 49, 50, 55, 57, 65, 66

context-driven image analysis 35, 45, 46, 65, 209

context navigation 44, 45, 182

convolutional layer 91, 93

convolutional neural network 91, 95, 96, 98

co-registration 52, 67, 105, 108–110, 122, 130, 138, 140, 181, 182, 197, 210

counterstain 14, 26, 52, 53

cross-validation 8, 75, 76, 102, 112, 120, 128, 145, 146, 210

CTLA-4 166, 200

cut-off 26

cytokine secretion 142

cytometry 12, 123, 163, 166

cytotoxic T cells   19, 104, 160, 200, 203

datafication   7–9, 21, 29, 166
data integration   7, 123, 124, 210
data management   26, 150, 197
data mining   3–5, 11, 31, 36, 101, 102, 104, 110–112, 122–130
decision support system (DSS)   4, 180, 182, 206, 211
decision tree   77, 78, 82, 212
deep learning   7, 22, 31, 39, 69, 71, 74, 87, 94, 97, 98, 100, 128, 195, 198
Definiens Developer XD   18, 27, 193
Definiens Result Containers   110
descriptive features   20
detection   13–16, 23, 24, 38, 39, 70, 71, 73, 83–86, 94–96, 99
development environment   18
diagnostic assays   10, 182
diagnostic knowledge   102
diagnostic pathology   175, 176, 178, 180, 182, 184
diagnostic readouts   11
digital analysis   37, 54, 150
digital pathology   51, 52, 82, 83, 144, 175–178, 180, 185–188, 190–193, 195–198
digitization   3, 4, 6, 53, 103, 182, 183, 186, 187, 189, 190, 200, 211, 212
disease biology   136–138
disease characteristics   21
disease prognosis   150
disease progression   102, 113–116, 118, 120–123, 127, 128, 151, 152, 178, 179, 210, 213
disruptive technology   186
distance metrics   164
DNA   1, 33, 132–136, 144, 151, 155

domain concept   20, 44
dropout layers   89, 93, 94
drug development   162, 199, 204, 207
DSS *see* decision support system
dual expression   13
dynamic imaging   188

early adopters   187
eCognition   46
endogenous pigment   13
epithelium   18, 140, 158, 164, 167, 170
ER *see* estrogen receptor
estrogen receptor (ER)   109, 114, 126, 148–151, 165, 179, 180
evolutionary approach   36
expected variation   24
experimental design   12–14
expert knowledge   21, 41, 43, 128, 177, 212
expression   13, 19–21, 114, 130, 135, 143, 148, 149, 152, 155, 165, 176, 200

F1 score   29, 30
false-color representation   107
FDA *see* US Food and Drug Administration
feature   70, 71, 77–81, 90, 91, 93, 111–113, 115–122, 128, 129, 211
feature learning   71, 77, 79, 81, 83, 85
feature selection   8, 102, 115, 117, 119–121, 129, 209, 211
feature validation   120, 121
FISH *see* fluorescent *in situ* hybridization
fixation time   14
flow cytometry   123, 166
fluorescent *in situ* hybridization (FISH)   132, 151

FOXP3 105, 108, 111, 142, 168, 173
fractality 20
fuzzy logic 43, 44

gabor features 81, 82
gene 2, 10, 11, 114, 134, 135, 137, 143, 144, 147, 148, 149, 154, 155, 165
gene expression 2, 10, 114, 135, 143, 148, 149, 151, 163, 165, 188, 209
gene expression analysis 149, 163
gene set variation analysis 137, 154
genetic algorithms 36
genetic methods 2
genetic mutations 2
genome 2, 3, 8, 70, 132 136, 155, 167, 172, 188, 202
genomic data 135, 143, 144, 146, 147, 150, 151, 166
genomics 2, 8, 10, 11, 115, 134, 135, 137, 139, 143–145, 147–152
gland 50, 98, 99, 159–161, 171, 173
glass slide 113, 180
gleason grading system 178
gold standard 1, 201, 205
grammatical concept 20
graphical language 18
ground truth data 23, 123
growth pattern 21

haar-like features 81, 82, 84
H&E *see* hematoxylin and eosin
H&E analysis problems 53
heatmap 106–111, 119, 122, 139, 140, 161, 213
  mining 140
hematoxylin 14, 26, 50, 52, 73, 104

hematoxylin and eosin (H&E) 50, 52–54, 58, 61, 65, 73, 83–85, 104, 105, 108, 109, 114, 165, 169
HER2 105, 112, 114, 126, 147–151, 179, 180, 192, 203
HER2 score 105
Herceptin treatment 150
hierarchical concept 20
hierarchical network 20, 43–45
hierarchical relations 43
hierarchy 19–21, 23, 42–45, 49, 71, 86, 177
high-throughput data 134, 145
high-throughput sequencing methods 134
histological section 1, 9, 73
histological slides 138, 158
histology 5, 15, 21, 50, 94, 99, 114, 150, 189
histopathology 1, 2, 6, 67, 70, 74, 98, 99, 102, 104, 184–187, 190, 192, 193
hormone therapy 150
hotspot detection 110
human genome project 132, 134
hybrid strategy 23
hypothesis-driven method 118, 119, 159,
hypothesis testing 192

IHC *see* immunohistochemistry
IM *see* invasive margin
image analysis 9–16, 18, 19, 23–28, 33–35, 38, 46, 65, 66, 121–124, 192, 193
image analysis software 11, 105, 180, 192, 194
image content information 177
image mining 7, 101, 102, 104, 124–126, 138, 140, 182, 211
image object domain 20
image object hierarchy 19, 20

image processing 67, 99, 118, 128, 138, 144, 166, 192, 197, 211
imaging artifact 194
imaging data 143, 146, 150
imaging mass cytometry 12, 163, 166
immune escape 158
immune landscape 157, 160, 166, 173, 179
immune oncology 158, 167
immune response 20, 142, 200, 203, 204, 208, 209
immune status 155, 157, 164, 200
immune therapy 169
immunofluorescence 12, 13, 95, 96, 163, 168, 187–189, 193, 196, 197
immunogenicity 158
immunohistochemistry (IHC) 2, 11–13, 50–53, 105, 114, 130, 132, 143, 147–152, 154, 168, 169, 179, 184, 190, 203, 211
immunoprofiling 163, 171
immunoscore 112, 142, 153, 162, 163, 170, 172, 179
immunosurveillance 163
immunotherapy 5, 159, 163, 166, 168, 200, 201, 209
Industry 4.0 3–5
infiltrating immune cells 164
inflammation 56, 157
input data 14, 26, 70, 72, 74, 80, 94, 212
in situ hybridization 2, 104, 132, 203
interactive scripting 18, 209
interferon gamma 151
internet 190
interobserver variability 160
inter-pathologist comparison 27
intra-epithelial tumor-infiltrating lymphocytes 20

invasive margin (IM) 16, 18, 20, 23, 32, 112, 140, 163, 179, 195, 212

Kaplan–Meier survival analysis 114
KI67 21, 104–110, 112, 168, 169, 179, 192
KI67 proliferation 21, 105

labeling 41, 194, 195, 197
LAG-3 166
limited clinical datasets 23
local processing 20, 65
local rules 21
loss function 91, 92
lymphocyte 17, 21, 22, 61, 62, 164, 170, 172
lymphocyte marker 21, 22

machine learning 22, 23, 39, 42, 65, 69–72, 74, 93, 94, 96–98, 129, 145, 181, 195, 196, 211, 212
machine learning approach 18, 22, 23, 69, 70, 74, 97, 98, 128, 158, 211, 212
macrophage 19, 164, 166, 171, 174
magnetic resonance imaging 104
map 16, 45, 48, 62, 84–86, 93, 107, 122
mass spectrometry 166
Medicine 4.0 3–5, 212
membership functions 21
meta-analysis 31, 142, 153
microscope 6, 177, 178, 180, 182, 188, 190
microsatellite instability high tumor 10
mobile phone 191
mobile scanning 191

molecular genetics 190
molecular network 3
molecular pathology 187
morphological assessment 2
morphological information 160
morphologic criteria 21
morphologic features 166
morphological observation 187
multiplex 32, 130, 139, 179
multiplexed analysis 163
multiplexed data 108–110
multiplexed immunofluorescence
    163, 187, 188, 193, 197
multiscale datasets 188
mutational status 119, 147

negative predictive value (NPV)
    116
neighboring objects 21, 47
neighborhood relationships 164
neoplastic cell 164
network of objects 17, 39, 43
next-generation sequencing 2,
    114, 131, 154, 155
non-small cell lung cancer (NSCLC)
    10, 32, 151, 154, 155, 168, 170
normalization 74, 114, 117, 118,
    128, 169, 194
NPV *see* negative predictive value
NSCLC *see* non-small cell lung
    cancer
nucleolus 20, 59
nucleus 13, 16, 17, 20, 39, 47, 50,
    55–58, 61, 62, 71, 94

object-based image analysis 42,
    46
object-based processing 43, 46,
    66
objective response rate 114
object primitives 17
objects of interest 17, 21, 45, 47,
    70, 83, 106, 111

observer variability 192, 204
omics 113, 123, 131, 133, 135,
    145, 154, 189, 196, 197, 212,
    213
Oncomine Dx Target Test 10
on-the-fly learning 83
optimization method 102
overall survival 114, 154, 159,
    167, 170
overfitting 73, 75, 79, 80, 93, 102,
    112, 128, 145, 195, 210

p63 160, 161
pancreatic cancer 142, 153
pan-genomic profiles 132
pathogenesis 11, 131, 149, 150
pathological assessment 23
pathologist annotation 16, 27,
    105
Pathology 4.0 4, 175, 182, 183,
    212
pathology data 144, 150
pathology lab 190
pathway activation 165, 174
pathways 51, 119, 132, 134, 136,
    137
patient care 1, 162, 182
patient cohort 116, 121
patient outcome 2
patient stratification 31, 150, 159
pattern recognition 69, 99, 100,
    144, 189–191, 195, 212
PD-L1 19, 32, 34, 65, 105, 151,
    152, 154, 155, 159, 200, 203
perceptron 87, 90–92, 100
performance metrics 115, 116
permutation test 121, 129, 145,
    146
personalized medicine 10, 143,
    164, 201, 203
personalized pathology 196
pharmaceutical industry 187,
    189, 199, 200

phene 2, 8, 113, 115–117, 121, 123, 128, 142, 158, 160, 182, 211, 213
phenome 3
phenotyping 3, 17, 157
point mutations 150, 205
point of care 188
polymerase chain reaction 132
positive percent agreement 30
positive predictive value (PPV) 30, 116
postgraduate training 190
PPV *see* positive predictive value
PR *see* progesterone receptor
precision medicine 10, 12, 138, 153, 183
predefined network architecture 22
prediction 69, 74, 76, 79–85, 96, 112, 116, 120, 121, 190, 191
predictive pathology 186
predictive value 77, 101, 112, 114–117, 120, 121, 196, 200, 203
prevalence 115–117, 120, 211
primary diagnosis 186–188
primitives 17, 23
process hierarchy 43–45
production environment 24
progesterone receptor (PR) 109, 114, 148–151, 179, 189
prognosis 142, 143, 145, 147, 148, 150, 151, 153, 154, 172, 173, 177, 178, 190, 191, 198, 199
prognostic features 8, 165, 167
prognostic influence 142, 153
prognostic model 167
prognostic value 142, 160, 172
progression free survival 159
proliferating cells 104
proportional proximity score 139

prostate cancer 113, 154, 159, 168, 171, 172, 178, 179, 184
proteomics 3, 212
proximity analysis 12
proximity measures 21

quality and control processes 140
quality assurance 176, 182
quality control 15, 25, 27, 29, 124, 125, 190

radiology 37, 104, 187, 190
radiomics 7, 8, 114, 115, 176, 212
random forests 70, 71, 77, 79–85, 88, 93, 97, 98, 212
randomized k-d-trees 125
readout 26
real-world images 36, 37
receptor status 150, 154
regions of interest 15, 70, 103, 104, 140, 160, 176, 181, 186, 195
region-specific scores 21
regression 33, 72, 74, 77–80, 84, 85, 99
regulatory approval 185–187
regulatory framework 177, 183
regulatory requirements 8
remote viewing 187
reproducibility 11, 33, 179, 192
RNA 1, 104, 125, 132, 134, 144, 151, 153, 154, 156, 173
ruleset 24

sample handling 14
scalable solutions 22
scanning artifacts 26
scene statistics 21
score 11, 29, 30, 77, 105, 112, 137, 139, 142, 159, 163, 167, 179, 180, 184

scoring 13, 24, 25, 112, 137, 157, 159, 164, 176, 179, 182, 192
second opinions 188
sectioning 14, 105
segmentation 16–19, 37, 46, 47, 55–60, 62, 73, 95, 99
selection bias 195, 206
semantic approach 18
semantics 39, 40–43
sensitivity 26, 76, 150–152
serial sections 13, 105, 106, 109, 181
signatures 11, 127, 137, 157, 158, 176, 213
slide-specific machine learning 23
social media 190
social network of cells 4–6, 66
Society 4.0 4
spatial distribution 139, 142
spatial interaction 73, 95, 153, 163, 185, 193
spatial localization 188
spatial relationship 142, 193, 209
staining gradients 21
staining quality 22, 26
staining variables 13
standardized quantification 192
statistical model 70, 145
streaming data sources 137
stroma 16, 18–20, 58, 70, 73, 140, 160, 161, 167, 181, 183, 196
study design 24
substratification 160
subvisual patterns 20
supervised learning 72, 73, 76, 212
survival time prediction 101
system 5, 6, 20, 41, 93, 109, 123–125, 167, 169, 178, 180, 190, 209–211
systems biology 191

tabular data outputs 30
tagging 41
technology infrastructure 186, 190
telepathology 183, 187, 188, 190, 198
tertiary lymphoid structures 20
test data 14, 26, 93, 111
texture 114, 195
The Cancer Genome Atlas 70, 136, 155
therapeutic option 10, 116
TILs *see* tumor-infiltrating lymphocytes
TIM-3 166
tissue biomarker 175, 213
tissue heatmap 213
tissue pathology 132
tissue phene 8, 142, 213
tissue phenomics 1–4, 6–12, 123–125, 157, 158, 169, 170, 174, 175, 177, 178, 180, 182, 212, 213
tissue preparation 21
tissue section 105, 139, 141, 161, 177, 179, 187, 188, 192, 204, 212, 213
tissue vascularization 160
TME *see* tumor microenvironment
toxicological screening 189
traditional histopathology 185, 186
training set 23, 24, 74, 75, 79
transcriptome screening 143
transcriptomic analysis 135–137, 142, 143, 149,
transcriptomics 3, 131, 150, 212
traveling salesman problem 36, 37
treatment decisions 1, 7, 9, 15, 35, 157, 167, 175, 179, 185, 199, 201, 203, 204, 206, 207

treatment selection 154, 157
tumor 10, 11, 16, 18, 20, 21, 95,
    96, 112, 113, 140–143, 149,
    150, 157–161, 163, 164, 193,
    194, 213, 214
  cells 13, 30, 104, 106, 108, 110,
    112, 142, 163, 164, 200, 205,
    212
  center 16, 18, 20, 23, 112, 140,
    141, 179
  environment 157
  heterogeneity 113, 126, 179,
    204
  histology 150
  microenvironment (TME) 6,
    9, 20, 32, 112, 113, 131, 138,
    140, 143, 157–162, 164, 166,
    171, 200, 203, 210, 213, 214
  region 16, 21, 52, 56, 95, 96
  stroma 18, 20
  surveys 24
tumor-infiltrating lymphocytes
    (TILs) 20, 21, 140, 142, 213

unsupervised cluster analysis
    102, 117
US Food and Drug Administration
    (FDA) 10, 127, 150, 176, 180,
    188, 191, 197

variability 1, 24– 26, 74, 83, 96,
    160, 192, 204
VeriTrova 27
virtual multiplexing 65, 181, 182
visual context 70, 71, 73, 77, 81,
    83–85
visualization 107, 108, 111, 127,
    129, 141, 184, 209, 211

whole-slide image (WSI) 94, 95,
    107–111, 184, 197, 198
whole-slide scanners 186, 188,
    192
workflow management 123–125
WSI *see* whole-slide image